S0-AZV-945

Wireless Sensor Networks

The Morgan Kaufmann Series in Networking
Series Editor, David Clark, M.I.T.

For further information on these books and for a list of forthcoming titles, please visit our website at http:www.mkp.com

Wireless Sensor Networks:
An Information Processing Approach

Feng Zhao
Microsoft Corporation

Leonidas J. Guibas
Stanford University

AMSTERDAM • BOSTON • HEIDELBERG • LONDON
NEW YORK • OXFORD • PARIS • SAN DIEGO
SAN FRANCISCO • SINGAPORE • SYDNEY • TOKYO
Morgan Kaufmann Publishers is an imprint of Elsevier

Senior Editor	Rick Adams
Associate Editor	Karyn Johnson
Publishing Services Manager	Simon Crump
Project Manager	Kyle Sarofeen
Cover Design	Dick Hannus Design Associates
Cover Image	Copyright © Dynamic Graphics, Inc.
Composition	Cepha Imaging Pvt. Ltd.
Technical Illustration	GGS Book Services
Copyeditor	Carol Leyba
Proofreader	Jacqui Brownstein
Indexer	Michael Ferreira
Interior printer	Maple-Vail Book Manufacturing Group
Cover printer	Phoenix Color Corp.

Morgan Kaufmann Publishers is an imprint of Elsevier.
500 Sansome Street, Suite 400, San Francisco, CA 94111

This book is printed on acid-free paper.

© 2004 by Elsevier Inc. All rights reserved.

Designations used by companies to distinguish their products are often claimed as trademarks or registered trademarks. In all instances in which Morgan Kaufmann Publishers is aware of a claim, the product names appear in initial capital or all capital letters. Readers, however, should contact the appropriate companies for more complete information regarding trademarks and registration.

No part of this publication may be reproduced, stored in a retrieval system, or transmitted in any form or by any means—electronic, mechanical, photocopying, scanning, or otherwise—without prior written permission of the publisher.

Permissions may be sought directly from Elsevier's Science & Technology Rights Department in Oxford, UK: phone: (+44) 1865 843830, fax: (+44) 1865 853333, e-mail: *permissions@elsevier.com.uk*. You may also complete your request on-line via the Elsevier homepage (*http://elsevier.com*) by selecting "Customer Support" and then "Obtaining Permissions."

Library of Congress Cataloging-in-Publication Data
Application submitted.

ISBN: 1-55860-914-8

For information on all Morgan Kaufmann publications,
visit our Web site at *www.mkp.com*.

Printed in the United States of America
    05 06 07 08   5 4 3 2

Contents

7

Sensor Network Platforms and Tools 239

Appendix

Preface

Wireless sensor networks have recently come into prominence because they hold the potential to revolutionize many segments of our economy and life, from environmental monitoring and conservation, to manufacturing and business asset management, to automation in the transportation and health-care industries. The design, implementation, and operation of a sensor network requires the confluence of many disciplines, including signal processing, networking and protocols, embedded systems, information management, and distributed algorithms. Such networks are often deployed in resource-constrained environments, for instance with battery operated nodes running untethered. These constraints dictate that sensor network problems are best approached in a holistic manner, by jointly considering the physical, networking, and application layers and making major design trade-offs across the layers. Consequently, for an emerging field such as sensor networks that involves a variety of different technologies, a student or practitioner often has to be versed in several disparate research areas before he or she can start to make contributions.

This book aims to provide a succinct introduction to the field of wireless sensor networks by synthesizing the diverse literature on key elements of sensor network design, such as sensor information organization, querying, and routing. The unifying theme throughout is the high-level information processing tasks that these networks are tasked to perform. It is our hope that this book will educate readers about the fundamental design principles and technology constraints of sensor networks, expose them to the many exciting and open research problems still present, and prepare them for new developments in this nascent area.

Audience

This book is intended for industry researchers and developers interested in sensor network systems and applications, and students who wish to pursue research in the sensor network field. It is also suitable for other professionals who desire to obtain an overview of the emerging sensor networks field. The book will introduce students and practitioners to the current, diverse research on sensor networks in a comprehensive manner, and expose the fundamental issues in designing and analyzing sensor network information processing systems. If used in an academic setting, the material in this book is suitable for a one-semester/quarter graduate or upper undergraduate level course. While the book strives to be self-contained, a number of chapters require some rudimentary knowledge of linear algebra, elementary probability and estimation theory, graph theory, networking protocols, databases, and distributed systems.

Approach

Writing a book for a field as dynamic and rapidly evolving as sensor networks is especially challenging. A first challenge is how to integrate techniques from a variety of disciplines that come into play in supporting high-level sensor network information processing tasks. We already mentioned some—and many more can be added to round out the list: signal processing and estimation, communication theory and protocols, distributed algorithms, probabilistic reasoning, databases, systems and software architectures, energy-aware computing, design methodologies, and evaluation metrics. What ties these different technologies together is the information processing demands of the task to be accomplished. The central theme of this book is to illustrate how information processing needs dictate, and are served by, the architecture of the different layers in a sensor network, including the establishment of services, the communication subsystem, sensor tasking and control, data management, and software design. Information processing provides the glue that allows us to integrate all this material together into a cohesive whole.

Within this large framework, however, a second challenge arises — how to remain concrete and focused, so as to best convey the material. We have chosen the problem of localizing and tracking moving targets as a canonical example to be used throughout. This is an example specific enough to be easy to grasp, yet general enough to motivate many fundamental sensor network issues, such as network discovery, service establishment, data routing and aggregation, query processing, and system organization, as well as trade-offs among them. By discussing the different layers of a sensor network within the context of a particular application, we are able to show how these layers collectively serve the needs of the application. This common problem also allows us to compare and evaluate different design approaches using a uniform set of metrics.

In summary, we have followed two principles in developing this book: stay with the information processing fundamentals and be concrete with examples. Much has been covered of the recent remarkable accomplishments in developing architectures for sensor networks, but in the underlying technologies a lot also remains subject to change. As the individual chapters make clear, sensor networks are a new, vibrant, and quickly evolving area. Even though the supporting hardware and software infrastructure may be subject to rapid evolution, it is the authors' belief that the information processing principles presented in this book will remain valid for a long time to come.

How to Use This Book

The material in the book can be covered in several different ways. Readers, however, should be aware of dependencies between some of the chapters. Chapters 1 and 2 introduce the sensor network field and the canonical problems used in the book—they should be covered first. Those who just want to get an overview of the field may skip Chapters 3–7, and go directly to Chapter 8 for an overview of emerging applications and future research directions. For those interested in the technical development, next should be Chapter 3 that introduces basic networking protocols and algorithms, as adapted for sensor networks. After that, Chapters 4 (infrastructure services), 5 (sensor tasking), and 6 (databases) may be covered in any order. We note that the material in Section 7.5 of Chapter 7 (platforms) depends on sensor group management and should be covered after Chapter 5 (sensor tasking).

Acknowledgments

This book arose out of a course—CS428: Information Processing for Sensor Networks—the authors taught at Stanford University during the spring of the 2001-02 and 2002-03 academic years. The authors wish to thank many individuals, without whose contributions this book would not have been possible.

- Students in the CS428 class helped the authors organize and debug the presentation of the materials that became the basis for this book.

- Jie Liu contributed Chapter 7 on sensor network platforms and tools. Qing Fang contributed the MAC section in Chapter 3. Patrick Cheung, Maurice Chu, Julia Liu, and Jim Reich helped with the preparation of the graphics used in the book.

- Parts of this book are based on research results obtained in joint work with Maurice Chu, Jie Gao, Horst Haussecker, Jie Liu, Julia Liu, Jim Reich, Jaewon Shin, and Feng Xie.

- PARC created a unique intellectual environment, within which this book was made possible. Other members of the PARC

CoSense Project, Patrick Cheung, Maurice Chu, Horst Haussecker, Qingfeng Huang, Xenofon Koutsoukos, Jim Kurien, Dan Larner, Jie Liu, Julia Liu, and Jim Reich provided stimulating discussions. Johan de Kleer, Andy Berlin, and John Gilbert advocated the Smart Matter systems research, from which the PARC sensor networks endeavor started.

- The Stanford Geometry group and its seminars created a lively environment for discussion of many of the topics in this book. Several members of the group directly contributed to improving the material of various chapters. Special thanks are due to Qing Fang, Jie Gao, Jaewon Shin, Feng Xie, and Danny Yang.

- Reviewers provided helpful comments and suggestions that significantly improved the organization and presentation of this book. These reviewers include Andrew Chien, Elaine Chong, Horst Haussecker, Jennifer Hou, Xenofon Koutsoukos, Bhaskar Krishnamachari, Zhen Liu, Shozo Mori, and John Stankovic. Also helpful were the reviewers during the proposal stage: Scott Bloom, John Gilbert, Horst Haussecker, Zhen Liu, Norm Whitaker, and Lixia Zhang.

- The editors of this book, Rick Adams and Karyn Johnson, of Morgan Kaufmann kept the writing on schedule and provided many aspects of editorial support.

The support of the Defense Advanced Research Projects Agency (DARPA) through the CoSense project at PARC and the Office of Naval Research (ONR) under a MURI grant to Stanford is gratefully acknowledged. Additional support was provided by PARC, Inc., and Honda Americas, Inc.

Feng Zhao
Leonidas J. Guibas

Palo Alto, California
January 2004

1

Introduction

The most profound technologies are those that disappear. They weave themselves into the fabric of everyday life until they are indistinguishable from it.

—The late Mark Weiser, Father of Ubiquitous Computing and
Chief Technologist of Xerox PARC

Advances in wireless networking, micro-fabrication and integration (for example, sensors and actuators manufactured using micro-electromechanical system technology, or MEMS), and embedded microprocessors have enabled a new generation of massive-scale sensor networks suitable for a range of commercial and military applications. The technology promises to revolutionize the way we live, work, and interact with the physical environment [222, 168, 65]. In the not-too-distant future, tiny, dirt-cheap sensors may be literally sprayed onto roads, walls, or machines, creating a digital skin that senses a variety of physical phenomena of interest: monitor pedestrian or vehicular traffic in human-aware environments and intelligent transportation grids, report wildlife habitat conditions for environmental conservation, detect forest fires to aid rapid emergency response, and track job flows and supply chains in smart factories. Unlike current information services such as those on the Internet where information can easily get stale or be useless because it is too generic, sensor networks promise to couple end users directly to sensor measurements and provide information that is precisely localized in time and/or space, according to the user's needs or demands.

With such technological advances come new challenges for information processing in sensor networks. What is needed are novel computational representations, algorithms and protocols, and design methodologies and tools to support distributed signal processing, information storage and management, networking, and application development. While this book will primarily focus on wireless sensor networks, some of the principles, such as those of collaborative information processing and management, apply equally well to wireline sensor networks. The issues of scalability and efficient use of bandwidth, a main concern of the book, are common to both wireless and wireline sensor networks.

1.1 Unique Constraints and Challenges

Unlike a centralized system, a sensor network is subject to a unique set of resource constraints such as finite on-board battery power and limited network communication bandwidth. In a typical sensor network, each sensor node operates untethered and has a microprocessor and a small amount of memory for signal processing and task scheduling. Each node is also equipped with one or more sensing devices such as acoustic microphone arrays, video or still cameras, infrared (IR), seismic, or magnetic sensors. Each sensor node communicates wirelessly with a few other local nodes within its radio communication range.

Sensor networks extend the existing Internet deep into the physical environment. The resulting new network is orders of magnitude more expansive and dynamic than the current TCP/IP network and is creating entirely new types of traffic that are quite different from what one finds on the Internet now. Information collected by and transmitted on a sensor network describes conditions of physical environments—for example, temperature, humidity, or vibration—and requires advanced query interfaces and search engines to effectively support user-level functions. Sensor networks may internetwork with an IP core network via a number of gateways, as in Figure 1.1. A gateway routes user queries or commands to appropriate nodes in a sensor network. It also routes sensor data, at times aggregated and summarized, to users who have requested it or are expected to utilize the information. A data repository or storage service may be present at the gateway, in addition to data logging at each sensor.

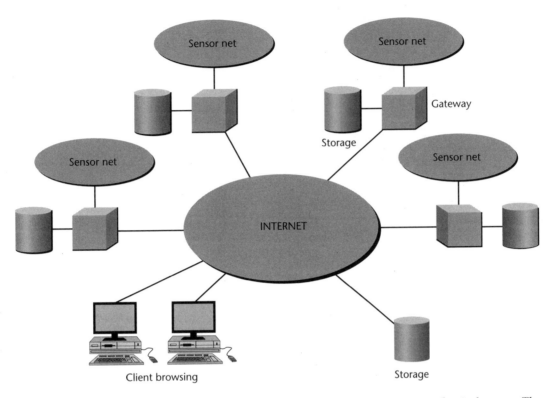

Figure 1.1 Sensor networks significantly expand the existing Internet into physical spaces. The data processing, storage, transport, querying, as well as the internetworking between the TCP/IP and sensor networks present a number of interesting research challenges that must be addressed from a multidisciplinary, cross-layer perspective.

The repository may serve as an intermediary between users and sensors, providing a persistent data storage. Additionally, one or more data storage devices may be attached to the IP network, to archive sensor data from a number of edge sensor networks and to support a variety of user-initiated browsing and search functions.

The current generation of wireless sensor hardware ranges from shoe-box-sized Sensoria WINS NG sensors [158] with an SH-4 microprocessor to matchbox-sized Berkeley motes with an 8-bit microcontroller [98]. A few samples of sensor hardware are shown in Figure 1.2; their corresponding capabilities are summarized and compared in Table 1.1. It is well known that communicating 1 bit over the wireless medium at short ranges consumes far more energy than

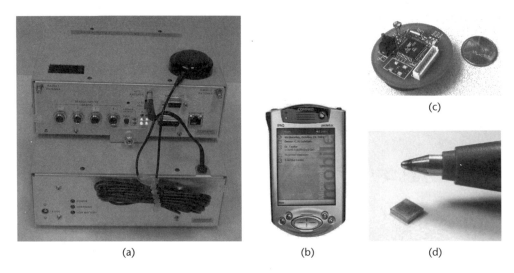

Figure 1.2 Samples of wireless sensor hardware: (a) Sensoria WINS NG 2.0 sensor node; (b) HP iPAQ with 802.11b and microphone; (c) Berkeley/Crossbow sensor mote, alongside a U.S. penny; (d) An early prototype of Smart Dust MEMS integrated sensor, being developed at UC Berkeley. (*Picture courtesy of Kris Pister and Jason Hill*).

processing that bit. For the Sensoria sensors and Berkeley motes, the ratio of energy consumption for communication and computation is in the range of 1000 to 10,000. Despite the advances in silicon fabrication technologies, wireless communication will continue to dominate the energy consumption of networked embedded systems for the foreseeable future [55].

Thus, minimizing the amount and range of communication as much as possible—for example, through local collaboration among sensors, duplicate data suppression, or invoking only the nodes that are relevant to a given task—can significantly prolong the life of a sensor network and leave nodes free to support multiuser operations. In addition, the shorter RF transmission range improves spectrum usage and increases throughput for a sensor network.

The information management and networking for this new network will require more than just building faster routers, switchers, and browsers. A sensor network is designed to collect information from a physical environment. Networking will be intimately coupled with the needs of sensing and control, and hence the application

Table 1.1 Comparison of the four sensor platforms shown in Figure 1.2.

	WINS NG 2.0 Node	iPAQ with 802.11 and A/D Cards in Sleeve	Berkeley MICA Mote*	Smart Dust**
Parts cost*** (quantity 1000+)	$100s	$100s	$10s	<$1
Size (cm^3)	5300	600	40	.002
Weight (g) (including battery)	5400	350	70	.002
Battery capacity (kJ)	300	35	15	(Less)
Sensors	Off-board	Microphone & light sensors integrated, others off-board	Integrated on PCB: Acceleration, temperature, light, sound	MEMS sensors to be integrated
Memory	32 MB RAM, 32 MB flash	64 MB RAM, 32 MB flash	4 KB RAM, 128 KB flash	(Less)
CPU	Hitachi SH4	StrongARM or XScale	ATmega 103L	(Less powerful)
Operating system	Linux	WinCE or Linux	TinyOS	(smaller)
Processing capability	400 MIPS/ 1.4 GFLOPS	240 MIPS	4 MIPS	(Less)
Radio range	100 m	100 m	30 m	(Shorter)

*The MICA mote is slightly larger than the WeC mote shown in Figure 1.2(c), and is more widely used.
**Smart Dust is not yet fully operational, but the size goal and power sources are known, and cost and weight are estimated.
***Note that the parts cost is based on large-quantity production.

semantics. To optimize for performance and resources such as energy, one has to rethink the existing TCP/IP stack and design an appropriate sensor network abstraction to support application development. For example, in many applications, it is more appropriate to address nodes in a sensor network by physical properties, such as node locations or proximity, than by IP addresses. How and where data is generated by sensors and consumed by users will affect the way data is compressed, routed, and aggregated. Because of the peer-to-peer connectivity and the lack of a global infrastructure support, the sensors have to rely on discovery protocols to construct local models about the network and environment. Mobility and instability in wireless links preclude the use of many existing edge-network gateway protocols for internetworking IP and sensor networks.

To summarize, the challenges we face in designing sensor network systems and applications include:

- *Limited hardware:* Each node has limited processing, storage, and communication capabilities, and limited energy supply and bandwidth.

- *Limited support for networking:* The network is peer-to-peer, with a mesh topology and dynamic, mobile, and unreliable connectivity. There are no universal routing protocols or central registry services. Each node acts both as a router and as an application host.

- *Limited support for software development:* The tasks are typically real-time and massively distributed, involve dynamic collaboration among nodes, and must handle multiple competing events. Global properties can be specified only via local instructions. Because of the coupling between applications and system layers, the software architecture must be codesigned with the information processing architecture.

1.2 Advantages of Sensor Networks

Networked sensing offers unique advantages over traditional centralized approaches. Dense networks of distributed communicating sensors can improve signal-to-noise ratio (SNR) by reducing average distances from sensor to source of signal, or target. Increased energy efficiency in communications is enabled by the multihop topology of the network [184]. Moreover, additional relevant information from other sensors can be aggregated during this multihop transmission through in-network processing [104]. But perhaps the greatest advantages of networked sensing are in improved robustness and scalability. A decentralized sensing system is inherently more robust against individual sensor node or link failures, because of redundancy in the network. Decentralized algorithms are also far more scalable in practical deployment and may be the only way to achieve the large scales needed for some applications.

1.2.1 Energy Advantage

Because of the unique attenuation characteristics of radio-frequency (RF) signals, a multihop RF network provides a significant energy saving over a single-hop network for the same distance. Consider the following simple example of an N-hop network. Assume the overall distance for transmission is Nr, where r is the one-hop distance. The minimum receiving power at a node for a given transmission error rate is $P_{receive}$, and the power at a transmission node is P_{send}. Then, the RF attenuation model near the ground is given by

$$P_{receive} \propto \frac{P_{send}}{r^\alpha},$$

where r is the transmission distance and α is the RF attenuation exponent. Due to multipath and other interference effects, α is typically in the range of 2 to 5. Equivalently,

$$P_{send} \propto r^\alpha P_{receive}.$$

Therefore, the power advantage of an N-hop transmission versus a single-hop transmission over the same distance Nr is

$$\eta_{rf} = \frac{P_{send(Nr)}}{N \cdot P_{send(r)}} = \frac{(Nr)^\alpha P_{receive}}{N \cdot r^\alpha P_{receive}} = N^{\alpha-1}. \tag{1.1}$$

Figure 1.3 illustrates the power attenuation for the multihop and single-hop networks. A larger N gives a larger power saving due to the consideration of RF energy alone. However, this analysis ignores the power usage by other components of an RF circuitry. Using more

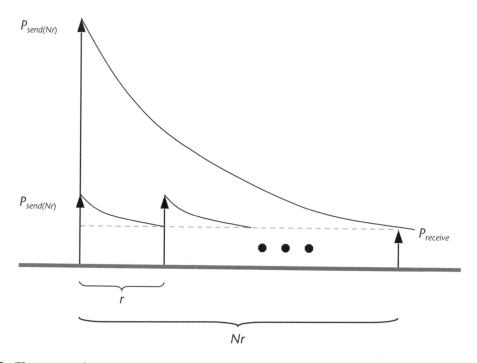

Figure 1.3   The power advantage of using a multihop RF communication over a distance of Nr.

nodes increases not only the cost, but also the power consumption of these other RF components. In practice, an optimal design seeks to balance the two conflicting factors for an overall cost and energy efficiency. Latency and robustness considerations may also argue against an unduly large number of relay nodes.

1.2.2 Detection Advantage

Each sensor has a finite sensing range, determined by the noise floor at the sensor. A denser sensor field improves the odds of detecting a signal source within the range. Once a signal source is inside the sensing range of a sensor, further increasing the sensor density decreases the average distance from a sensor to the signal source, hence improving the signal-to-noise ratio (SNR). Let us consider the acoustic sensing case in a two-dimensional plane, where the acoustic power received at a distance r is

$$P_{receive} \propto \frac{P_{source}}{r^2},$$

which assumes an inverse distance squared attenuation. The SNR is given by

$$\text{SNR}_r = 10 \log \frac{P_{receive}}{P_{noise}} = 10 \log P_{source} - 10 \log P_{noise} - 20 \log r.$$

Increasing the sensor density by a factor of k reduces the average distance to a target by a factor of $\frac{1}{\sqrt{k}}$. Thus, the SNR advantage of the denser sensor network is

$$\eta_{snr} = \text{SNR}_{\frac{r}{\sqrt{k}}} - \text{SNR}_r = 20 \log \frac{r}{\frac{r}{\sqrt{k}}} = 10 \log k. \qquad (1.2)$$

Therefore, an increase in sensor density by a factor of k improves the SNR at a sensor by $10 \log k$ db.

1.3 Sensor Network Applications

A sensor network is designed to perform a set of high-level information processing tasks such as detection, tracking, or classification. Measures of performance for these tasks are well defined, including detection of false alarms or misses, classification errors, and track quality. Applications of sensor networks are wide ranging and can vary significantly in application requirements, modes of deployment (e.g., ad hoc versus instrumented environment), sensing modality, or means of power supply (e.g., battery versus wall-socket). Sample commercial and military applications include:

- Environmental monitoring (e.g., traffic, habitat, security)

- Industrial sensing and diagnostics (e.g., appliances, factory, supply chains)

- Infrastructure protection (e.g., power grids, water distribution)

- Battlefield awareness (e.g., multitarget tracking)

- Context-aware computing (e.g., intelligent home, responsive environment)

Chapter 8 will examine the application space in greater detail. For now, we look at a few application samples in order to ground the technical discussions for the rest of the book.

1.3.1 Habitat Monitoring: Wildlife Conservation Through Autonomous, Nonintrusive Sensing

On a small patch of land 10 miles off the coast of Maine, a team of computer engineers from the University of California, Berkeley, are conducting an experiment in networked sensing. Working with biologists at the College of the Atlantic, the engineers have installed 190 wireless sensors that are being used to monitor the habitat of the nesting petrels[1] on Great Duck Island (see Figure 1.4) [150]. In the past, biologists studying the nesting behaviors of these birds had to travel to the island every now and then to gather observation data. To check on the petrels, they literally had to stick their hands into the burrows, often causing the birds to abandon their homes.

Now, these same biologists are checking on the birds on the island in the comfort of their offices, browsing data from the sensors linked by satellite. And their colleagues thousands of miles away can share the same experience, thanks to the Internet. The untethered, matchbox-sized sensors left in the burrows monitor the occupancy by recording temperature variations inside and wirelessly send the data to a gateway node on the island. Convenience aside, the more significant benefit of the technology is the minimization of disturbance to the very habitat that scientists are trying to help preserve.

---

1 This is the name given to various ocean birds belonging to the order of tube-nosed swimmers, like the albatross and the shearwater.

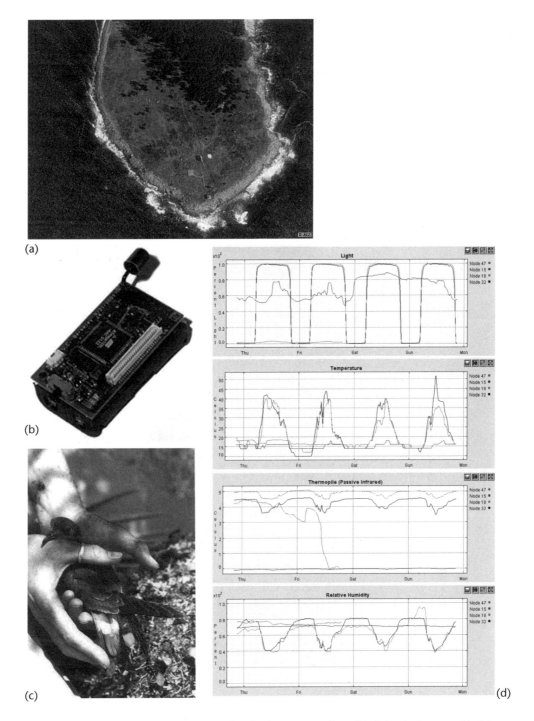

Figure 1.4 Remote wireless sensors are used to find out more about birds in their natural habitat. (a) Sensor network deployment on Great Duck Island, 10 miles off the coast of Maine. (b) Wireless biological sensors placed in bird nests. (c) A petrel, rarely seen by birdwatchers. (d) Sensor measurements remotely retrieved using a Web interface, summer 2002. (*Picture courtesy of U.S. Geological Survey, Intel Corporation, and John Anderson*).

The experiment on Great Duck Island is a small lens into an expansive future. To grasp what might happen, multiply these 190 sensors by 10 million or 100 million and distribute them globally. When the sensor grid becomes ubiquitous, an enormous digital retina will stretch over the surface of the planet. This planet-scale system could help us understand and address tomorrow's environmental challenges, ranging from monitoring global biodiversity to sensing millions of low-level, nonpoint sources of pollution [230].

1.3.2 Tracking Chemical Plumes: Ad Hoc, Just-in-Time Deployment for Mitigating Disasters

Image the following scenario. The Valley Authority has just declared a region-wide emergency: A large-scale hazardous chemical gas leak occurred at a chemical processing plant twenty minutes ago. The National Guard has been activated to evacuate nearby towns and to close roads and bridges. To get a real-time situational assessment of the extent and movement of the gas release and help plan the evacuation, the SensorNet SWAT Team is called in. Three unmanned aerial vehicles (UAVs) are immediately launched from an open field 15 miles south of the accident site, each carrying 1000 tiny wireless chemical sensing nodes (see Figure 1.5). Upon flying over the vicinity of the accident site, the sensor nodes are air-dropped. While airborne, the nodes self-organize into an ad hoc network and relay the sensing result back to the UAVs nearby: Where is the plume? How big is it? What is the shape? How fast is it moving?

While the full realization of this scenario may take several years, pieces of the sensor network technology have already been demonstrated. For example, in March 2001, Kris Pister and his Berkeley team deployed wireless sensors using a UAV for a vehicle tracking field demonstration in the Mojave Desert of Southern California. Each sensor node is equipped with a microprocessor, an RF modem, and a magnetometer. Including the battery, the sensor node weighs about an ounce. The UAV air-dropped six wireless sensors along a road (Figure 1.6). Once on the ground, the sensors immediately

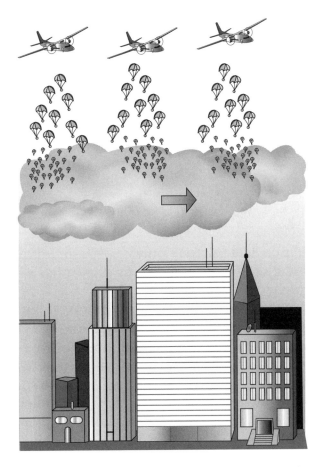

Figure 1.5 Tracking chemical plumes using ad hoc wireless sensors, deployed from air vehicles.

self-organized into a network, initiated detection when a vehicle passed by, and relayed the tracking results back to the UAV.

1.3.3 Smart Transportation: Networked Sensors Making Roads Safer and Less Congested

Plenty of sensors are already in use for traffic monitoring purposes. Sensors embedded in roadbeds or alongside highways measure traffic flow. Cameras at street intersections look for traffic violations.

Figure 1.6 Left: Berkeley wireless sensor mote. Right: Air-drop of six sensor nodes from a UAV. (*Picture courtesy of Kris Pister and Jason Hill.*)

Sensors in vehicles monitor speed and other conditions. But today these sensors do not talk to each other as often as we would like them to.

When these sensors are networked together to share real-time information, we can begin to create a dynamic infrastructure for smart roads that can be optimized to make roads safer, reduce congestion, or help people find the nearest available parking space in an unfamiliar city. For example, cars and trucks equipped with wireless sensors can warn each other of imminent collision or other road hazards ahead. They can dynamically optimize routes to avoid traffic hot spots. They can continuously monitor their own conditions and transmit emission data to a service station to enable just-in-time maintenance. They can use the wireless connection to download music and movies while waiting at a gas station. For some of these applications, the crucial difference lies in the real-time flow of information. Point-to-point routing assistance or road safety application requires access to up-to-date or even instantaneous traffic data.

When sensors are used to guard critical infrastructure such as power plants or airports, they can provide a virtual enclosure around the facilities to guard against unauthorized intrusion. Unlike existing security systems that require human personnel to watch video feeds around the clock and pick out "unusual" events, which is clearly not scalable, smart sensor networks will have to rely on in-network intelligence to focus on interesting events, filter out detractors, extract

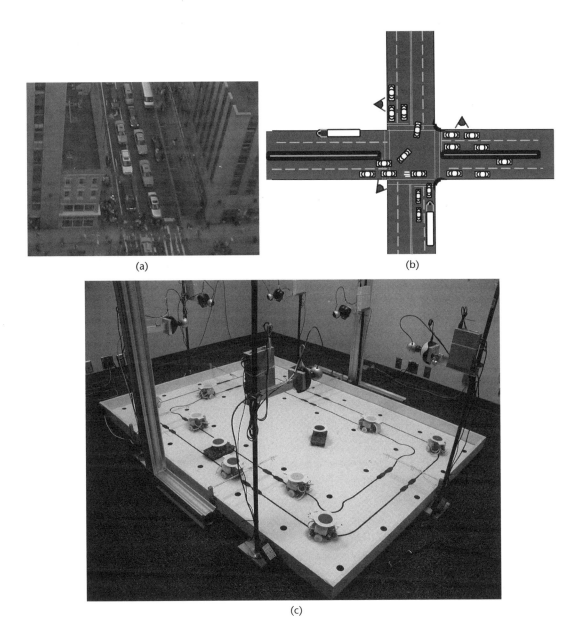

Figure 1.7 Distributed video sensor networks for traffic and security applications. Upper figures: Networked cameras and other sensors could be used to monitor traffic flow to reduce congestion, track vehicles on city streets for traffic violations, or detect illegal activities around critical infrastructure such as airports. Lower figure: PARC video sensor network prototype uses in-network intelligence to decide what events to pay attention to and what to ignore, thus reducing the amount of information the network must collect and transport in order to support high-level monitoring applications.

meaning from raw sensor readings or video streams, and transmit only relevant scene features to human users.

1.4 Collaborative Processing

As the above applications have illustrated, many sensing tasks require a sensor network system to process data cooperatively and to combine information from multiple sources. In traditional centralized sensing and signal processing systems, raw data collected by sensors are relayed to the edges of a network where the data is processed. From the scalability point of view, the nonlocal processing at the edges depletes precious bandwidth. If every sensor has some data that it needs to send to another node in a network, then a well-known wireless capacity result by Gupta and Kumar states that the per node throughput scales as $\frac{1}{\sqrt{N}}$—in other words, it goes to zero as the number of nodes N in a wireless ad hoc network increases [88]. This result holds regardless of optimality in routing, power control, or transmission. Intuitively, this states that, as the number of nodes increases, every node spends almost all of its time forwarding packets of other nodes. From the energy point of view, transmitting raw data to distant nodes is wasteful of scarce resources. The diminishing wireless capacity result can be somewhat mitigated by introducing mobility to nodes, if an application is delay-tolerant [85].

In a sensor network context, one can clearly do better. Since data from multiple sensors with overlapping sensing regions is almost always correlated, one can remove the redundant information in the data, through in-network aggregation and compression local to the nodes that generate the data, before shipping it to a remote node. In fact, it can be shown that the amount of nonredundant data that a network generates grows as $O(\log N)$, assuming that the network is sampling a physical phenomenon with a prescribed accuracy requirement [206]. This is encouraging since the amount of data generated per node scales as $O\left(\frac{\log N}{N}\right)$, which is within the per-node throughput constraint derived by Gupta and Kumar. On the other hand, instead of applying data compression techniques to samples after they are collected, nodes can be more selective in what data to generate or communicate. For energy-constrained and multiuser decentralized systems, it becomes critical to carefully select the sensor nodes that participate in a sensor collaboration, balancing the information contribution of each against its resource consumption or potential utility for other users. We use the term *collaborative signal and information processing* (CSIP) to refer to signal and information processing problems dominated by this issue of selecting embedded sensors to participate in an information processing task [232]. The topic of active control and tasking of sensors is quite unique to sensor networks and will be covered in greater details in Chapter 5.

1.5 Key Definitions of Sensor Networks

Sensor networks is an interdisciplinary research area that draws on contributions from signal processing, networking and protocols, databases and information management, distributed algorithms, and embedded systems and architecture. In the following, we define a number of key terms and concepts that will be used throughout the book as we develop techniques and examples for sensor networks.

- *Sensor:* A transducer that converts a physical phenomenon such as heat, light, sound, or motion into electrical or other signals that may be further manipulated by other apparatus.

- *Sensor node:* A basic unit in a sensor network, with on-board sensors, processor, memory, wireless modem, and power supply. It is often abbreviated as *node*. When a node has only a single sensor on board, the node is sometimes also referred to as a *sensor*, creating some confusion.

- *Network topology:* A connectivity graph where nodes are sensor nodes and edges are communication links. In a wireless network, the link represents a one-hop connection, and the neighbors of a node are those within the radio range of the node.

- *Routing:* The process of determining a network path from a packet source node to its destination.

- *Date-centric:* Approaches that name, route, or access a piece of data via properties, such as physical location, that are external to a communication network. This is to be contrasted with address-centric approaches which use logical properties of nodes related to the network structure.

- *Geographic routing:* Routing of data based on geographical attributes such as locations or regions. This is an example of date-centric networking.

- *In-network:* A style of processing in which the data is processed and combined near where the data is generated.

- *Collaborative processing:* Sensors cooperatively processing data from multiple sources in order to serve a high-level task. This typically requires communication among a set of nodes.

- *State:* A snapshot about a physical environment (e.g., the number of signal sources, their locations or spatial extent, speed of movement), or a snapshot of the system itself (e.g.,the network state).

- *Uncertainty:* A condition of the information caused by noise in sensor measurements, or lack of knowledge in models. The uncertainty affects the system's ability to estimate the state accurately and must be carefully modeled. Because of the ubiquity of uncertainty in the data, many sensor network estimation problems are cast in a statistical framework. For example, one may use a covariance matrix to characterize the uncertainty in a Gaussian-like process or more general probability distributions for non-Gaussian processes.

- *Task:* Either high-level system tasks which may include sensing, communication, processing, and resource allocation, or application tasks which may include detection, classification, localization, or tracking.

- *Detection:* The process of discovering the existence of a physical phenomenon. A threshold-based detector may flag a detection whenever the signature of a physical phenomenon is determined to be significant enough compared with the threshold.

- *Classification:* The assignment of class labels to a set of physical phenomena being observed.

- *Localization and tracking:* The estimation of the state of a physical entity such as a physical phenomenon or a sensor node from a set of measurements. Tracking produces a series of estimates over time.

- *Value of information or information utility:* A mapping of data to a scalar number, in the context of the overall system task and knowledge. For example, information utility of a piece of sensor data may be characterized by its relevance to an estimation task at hand and computed by a mutual information function.

- *Resource:* Resources include sensors, communication links, processors, on-board memory, and node energy reserves. Resource allocation assigns resources to tasks, typically optimizing some performance objective.

- *Sensor tasking:* The assignment of sensors to a particular task and the control of sensor state (e.g., on/off, pan/tilt) for accomplishing the task.

- *Node services:* Services such as time synchronization and node localization that enable applications to discover properties of a node and the nodes to organize themselves into a useful network.

- *Data storage:* Sensor information is stored, indexed, and accessed by applications. Storage may be local to the node where the data is generated, load-balanced across a network, or anchored at a few points (warehouses).

- *Embedded operating system (OS):* The run-time system support for sensor network applications. An embedded OS typically provides an abstraction of system resources and a set of utilities.

- *System performance goal:* The abstract characterization of system properties. Examples include scalability, robustness, and network longevity, each of which may be measured by a set of evaluation metrics.

- *Evaluation metric:* A measurable quantity that describes how well the system is performing on some absolute scale. Examples include packet loss (system), network dwell time (system), track loss (application), false alarm rate (application), probability of correct association (application), location error (application), or processing latency (application/system). An evaluation method is a process for comparing the value of applying the metrics on an experimental system with that of some other benchmark system.

1.6 The Rest of the Book

The rest of the book is organized to reflect the basic building blocks of a typical sensor network and its applications. Because of the diversity of sensor network applications, we believe it is important to ground the technical development of the book on a small number of carefully chosen examples. Tracking is such an example. Chapter 2 introduces the tracking problem, task requirements, a formulation of localization and tracking in a probabilistic framework, and important issues in designing information processing algorithms in sensor networks.

Fundamental to sensor network operations are the networking layer and the infrastructure services to support the correct functioning of nodes. Chapters 3 and 4 provide a systematic overview of the networking algorithms and protocols for sensor networks, including geographic, energy-aware, and attribute-based routing protocols, as well as algorithms for maintaining time synchronization and for discovering locations.

Unique to sensor networks is the ability to dynamically task sensors for data collection, processing, and communication. Chapter 5 introduces information-based approaches to sensor tasking, including models of utility and costs and algorithms for optimization of information gathering. Treating a sensor network as a distributed database, Chapter 6 introduces important concepts of distributed data indexing and access. The ability to quickly locate a piece of information at a modest cost is essential for many in-network processing tasks as well as for interfacing with end users.

To support rapid prototyping of applications for a sensor network, we must develop system support tools and a software environment that provide a high-level abstraction of a distributed system and at the same time expose important resource constraints to application developers. Chapter 7 introduces programming models and software tools suitable for these tasks. Finally, Chapter 8 wraps up the book by providing a systematic look at the application space for sensor networks and outlining a number of important research directions.

2

Canonical Problem: Localization and Tracking

Localizing and tracking moving stimuli or objects is an essential capability for a sensor network in many practical applications. Moreover, it is a familiar problem that can be used as a vehicle to study many information processing and organization problems for sensor networks. For example, as we mentioned in Section 1.4, a central problem for collaborative signal and information processing (CSIP) is to dynamically define and form sensor groups based on task requirements and resource availability. Tracking exposes the most important issues surrounding collaborative processing, information sharing, and group management including which nodes should sense, which have useful information and should communicate, which should receive the information and how often, and so on, all in a dynamically evolving environment.

From a sensing and information processing point of view, we define a sensor network as an abstract tuple, $G = \langle V, E, P_V, P_E \rangle$. V and E specify a network graph, with its nodes V, and link connectivity $E \subseteq V \times V$. P_V is a set of functions that characterizes the properties of each node in V, such as its location, computational capability, sensing modality, sensor output type, energy reserve, and so on. Possible sensing modalities include acoustic, seismic, magnetic, IR, temperature, or light. Possible types of sensor output include information about signal amplitude, source direction-of-arrival (DOA), target range, or target classification label. Similarly, P_E specifies properties for each link, such as link capacity and quality.

We also recognize that another class of sensor network applications is concerned with the problem of sensing a field. A field is a distributed physical quantity, such as temperature, pressure, or optical flow across a region of space. Each point in a field is associated with a scalar or vector value. A temperature field defines a temperature value for each point in a planar area or spatial volume—mathematically, it is a mapping $\mathcal{R}^2 \to \mathcal{R}^1$ or $\mathcal{R}^3 \to \mathcal{R}^1$. In a temperature field sensing problem, one may be interested in knowing the locations or spatial extent of hot spots. To answer such queries, one may first extract iso-contours of the temperature function from pointwise sensor measurements and then characterize these contours according to their size, or other factors. Although this may seem quite different from the problem of tracking point targets, both require collaborative processing among sensors. In the tracking case, the collaborative processing proceeds along the temporal dimension as well as in the spatial domain, especially when multiple targets interact. In the field sensing case, the collaboration among sensors primarily occurs in the spatial domain and occasionally along the temporal dimension when the field evolves over time. In the rest of this chapter, we will focus on the tracking problem, in order to bring out the key information processing issues.

A tracking task can be formulated as a constrained optimization problem $\langle G, T, W, Q, J, \mathcal{C} \rangle$. Here G is the sensor network specified earlier, T is a set of targets, where we specify for each target the location, shape (if not a point source), and signal source type, and W is a signal model for how target signals propagate and attenuate in the physical medium. For example, a possible power attenuation model for an acoustic signal is the inverse distance squared model. The variable Q denotes a set of user queries, specifying query instances and query entry points into the network. A sample query is "Count the number of targets in region R." The variable J specifies an objective function, defined by task requirements. For example, for a target localization task, the objective function could be the localization accuracy, expressed as the "size" of the covariance matrix for the position estimate. Finally, $\mathcal{C} = \{C_1, C_2, \dots, \}$ specifies a set of constraints. An example is localizing an object within a certain amount of time and using no more than a certain quantity of energy. For a constrained optimization problem, a solver finds a set of feasible sensing and communication solutions that satisfy the given set of constraints. For example, a solution to the just-cited localization problem could specify a set of sensor nodes on a path that gather and combine data as they route the result back to the querying node.

In wireless sensor networks, some of the information defining the objective function and constraints is available only at run time. Furthermore, the optimization problem may have to be solved in a decentralized way. Often anytime algorithms are desirable because constraints and resource availability may change dynamically. Consequently, the decentralized algorithms and protocols for solving the optimization problem are quite different from existing centralized optimization techniques.

2.1 A Tracking Scenario

We use the following tracking scenario to bring out key CSIP issues. As a target X moves from left to right across the sensor field, a number of activities are initiated in the network, as shown in Figure 2.1.

1. *Discovery*: Node a detects X and initializes tracking.

2. *Query processing*: A user query Q enters the network and is routed toward regions of interest—in this case, the region around node a. It should be noted that other types of queries, such as long-running queries that dwell in a network over a period of time, are also possible.

3. *Collaborative processing*: Node a estimates the target location, possibly with help from neighboring nodes. The position estimation may be accomplished by a triangulation or a least-squares computation over a set of sensor measurements. More generally, the estimate may be obtained using a statistical method such as Bayesian estimation, which is detailed in Section 2.2.

4. *Communication*: As the target X moves, node a may hand off an initial estimate of the target location to node b, b to c, and so on.

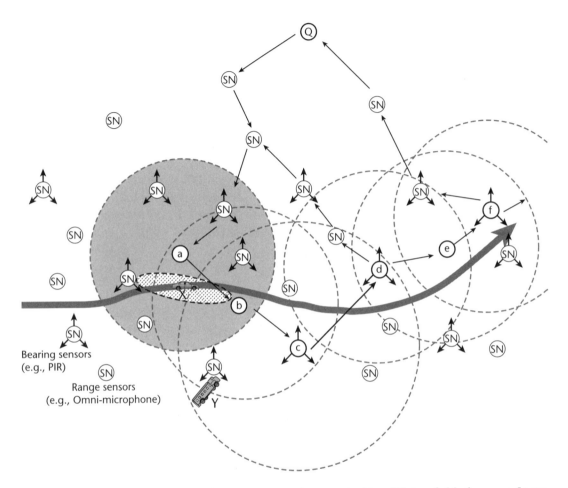

Figure 2.1   A tracking scenario, showing two moving targets, X and Y, in a field of sensors. Large dashed circles represent the range of radio communication for each node (adapted from [232]).

One of the key problems here is to select the next node among a's 1-hop neighbors. A poor choice of b may cause the network to lose the target or incur unnecessary communication overhead. The communication of data may be inseparable from data aggregation and processing and may have to be jointly optimized with processing.

5. *Reporting*: Node d or f may summarize track data and send it back to the querying node.

Assume now that another target, Y, enters the region around the same time. The network will have to handle multiple tasks in order to track both targets simultaneously. When the two targets move close to each other, the problem of properly associating a measurement to a target track, the so-called *data association* problem, has to be addressed. In addition, collaborative sensor groups, as defined earlier, must be selected carefully since multiple groups might need to share the same physical hardware [144].

This tracking scenario raises a number of fundamental information processing issues in distributed information discovery, representation, communication, storage, and querying:

- In collaborative processing, the issues of target detection, localization, tracking, and sensor tasking and control

- In networking, the issues of data naming, aggregation, and routing

- In databases, the issues of data abstraction and query optimization

- In human-computer interface, the issues of data browsing, search, and visualization

- In infrastructure services, the issues of network initialization and discovery, time and location services, fault management, and security

2.2 Problem Formulation

We use the following notation in our formulation of the tracking problem in a sensor network:

- Superscript t, where applicable, denotes time. We consider discrete times t that are nonnegative integers.

- Subscript $i \in \{1,\dots,K\}$, where applicable, denotes the sensor index; K is the total number of sensors in the network.

- Subscript $j \in \{1,\dots,N\}$, where applicable, denotes the target index; N is the total number of targets being observed.

- The target state at time t is denoted as $\mathbf{x}^{(t)}$. For a multitarget tracking problem, this is a concatenation of individual target states $\mathbf{x}_j^{(t)}$. Without loss of generality, we consider in this chapter the tracking problem, where an individual target state is the location of a moving point object in a two-dimensional plane.

- The measurement of sensor i at time t is denoted as $\mathbf{z}_i^{(t)}$. In the context of discussing estimation problems, we will use the terms *state* and *parameter* interchangeably.

- The measurement history up to time t is denoted as $\overline{\mathbf{z}^{(t)}}$, that is, $\overline{\mathbf{z}^{(t)}} = \left\{ \mathbf{z}^{(0)}, \mathbf{z}^{(1)}, \dots, \mathbf{z}^{(t)} \right\}$. The measurements may originate from a single sensor or a set of sensors.

- The collection of all sensor measurements at time t are denoted as $\underline{\mathbf{z}^{(t)}}$, that is, $\underline{\mathbf{z}^{(t)}} = \left\{ \mathbf{z}_1^{(t)}, \mathbf{z}_2^{(t)}, \dots, \mathbf{z}_K^{(t)} \right\}$.

- In general, bold-face lowercase symbols denote vector quantities such as position or velocity, while bold-face uppercase symbols denote matrices such as steering matrix used in direction-of-arrival (DOA) estimation.

2.2.1 Sensing Model

We will formulate our estimation problem using standard estimation theory. The time-dependent measurement, $\mathbf{z}_i^{(t)}$, of sensor i with characteristics $\lambda_i^{(t)}$ is related to the parameters, $\mathbf{x}^{(t)}$, that we wish to estimate through the following observation (or measurement) model,

$$\mathbf{z}_i^{(t)} = \mathbf{h}\left(\mathbf{x}^{(t)}, \lambda_i^{(t)} \right), \tag{2.1}$$

where $\mathbf{h}$ is a (possibly nonlinear) function depending on $\mathbf{x}^{(t)}$ and parameterized by $\lambda_i^{(t)}$, which represents our (possibly time-dependent) knowledge about sensor i. Here, we consider the sensing model for a single target, with $\mathbf{x}$ representing the location of the target. Typical characteristics, $\lambda_i^{(t)}$, about sensor i include sensing modality (which refers to what kind of sensor i is), sensor position ζ_i, and other parameters, such as the noise model of sensor i and its power reserve. Typically, the sensor characteristics are relatively stable. This explicit representation of the sensor characteristics allows us to separate the relatively stable knowledge about the sensors from the more dynamic measurements in order to optimize for the distributed estimation.

In (2.1), we consider a general form of the observation model that accounts for possibly nonlinear relations between the sensor type, sensor position, noise model, and the parameters we wish to estimate. A special case of (2.1) would be

$$\mathbf{h}\left(\mathbf{x}^{(t)}, \lambda_i^{(t)}\right) = \mathbf{f}_i\left(\mathbf{x}^{(t)}, \lambda_i^{(t)}\right) + \mathbf{w}_i^{(t)},$$

where $\mathbf{f}_i$ is a (possibly nonlinear) observation function, and $\mathbf{w}_i$ is additive, zero mean noise with known covariance.

In case $\mathbf{f}_i$ is a linear function on the parameters, (2.1) reduces to the linear equation

$$\mathbf{h}\left(\mathbf{x}^{(t)}, \lambda_i^{(t)}\right) = \mathbf{H}_i^{(t)}\left(\lambda_i^{(t)}\right)\mathbf{x}^{(t)} + \mathbf{w}_i^{(t)}. \tag{2.2}$$

In order to illustrate the idea, we consider the problem of stationary target localization with time-invariant sensor characteristics. Here, we assume that all sensors are acoustic sensors measuring only the amplitude of the sound signal so that the state vector $\mathbf{x} = [x, y]^T$ is the unknown target position, and

$$\lambda_i = \left[\zeta_i, \sigma_i^2\right]^T, \tag{2.3}$$

where ζ_i is the known sensor position and σ_i^2 is the known additive noise variance. Note there is no longer a time dependence for $\mathbf{x}$ and λ_i. Assuming that acoustic signals propagate isotropically, the parameters are related to the measurements by

$$z_i = \frac{a_i}{\|\mathbf{x} - \zeta_i\|^{\frac{\alpha}{2}}} + w_i, \qquad (2.4)$$

where a_i is a given random variable representing the amplitude of the signal at the target, α is a known attenuation coefficient, and $\|\cdot\|$ is the Euclidean norm. The term w_i is a zero mean Gaussian random variable with variance σ_i^2. The characteristics of the acoustic sensor is examined in greater details in Section 2.5.

2.2.2 Collaborative Localization

Consider the problem of localizing a stationary signal source using a set of sensor measurements. In the simplest setting, three or more amplitude measurements, say from microphones, may be used to determine the location of a signal source (see Figure 2.2). If the signal attenuation model is known [such as the acoustic model (2.4)], one can recover one range constraint per amplitude measurement. To uniquely determine the location on a two-dimensional plane, one needs at least three independent distance measurements (the third is needed to resolve ambiguities). Alternately, one may use time difference of arrival (TDOA) of signals at the sensors to estimate the range or bearing information. Various ranging techniques are discussed in Chapter 4. Here we focus on localization using signal amplitude measurements.

Assume $\alpha = 2$ in the signal propagation model (2.4), which is equivalent to the inverse distance squared model for power attenuation. Let $\mathbf{x} \in \mathcal{R}^2$ be the position of the signal source, $\zeta_i \in \mathcal{R}^2$ be the position of sensor i, and z_i be the amplitude measurement of the sensor. Omitting the noise term, we can rewrite the signal model (2.4) as

$$\|\mathbf{x}\|^2 + \|\zeta_i\|^2 - 2\mathbf{x}^T\zeta = \frac{a_i}{z_i}, \quad i = 1, 2, 3, \ldots$$

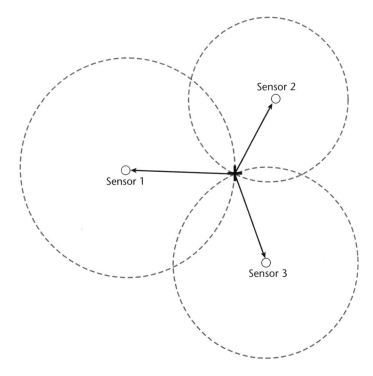

Figure 2.2 Localization: Three measurements are used to localize a signal source in a plane.

For each sensor i, this equation gives a quadratic constraint on the unknown $\mathbf{x}$. To generate a set of linear constraints from which we can solve for $\mathbf{x}$, we subtract the $i = 1$ constraint from the rest ($i \neq 1$) and obtain

$$-2(\zeta_i - \zeta_1)^T \mathbf{x} = a_i \left(\frac{1}{z_i} - \frac{1}{z_1} \right) - (\|\zeta_i\|^2 - \|\zeta_1\|^2).$$

Letting $c_i = -2(\zeta_i - \zeta_1)$ and $d_i = a_i \left(\frac{1}{z_i} - \frac{1}{z_1} \right) - (\|\zeta_i\|^2 - \|\zeta_1\|^2)$, we can simplify the above as

$$c_i^T \mathbf{x} = d_i.$$

Given K sensors, we can obtain $K-1$ linear constraints, expressed in the matrix form

$$\mathbf{C}_{K-1}\mathbf{x} = \mathbf{d}_{K-1}.$$

For $K = 3$, the above uniquely determines the location of the signal source $\mathbf{x}$. For $K > 3$, we can solve for it using the least-squares method [214]:

$$\mathbf{x} = \left[\left(\mathbf{C}_{K-1}^T\mathbf{C}_{K-1}\right)^{-1}\mathbf{C}_{K-1}^T\right]\mathbf{d}_{K-1}.$$

There are a few points worth discussing here. Not every constraint above provides the same amount of information for the localization task. The geometry of the sensor placement as well as the distance to the signal source (i.e., SNR) determine the significance of the contribution of each sensor. For example, if three sensors are collinear, they no longer uniquely determine the location of the signal source. It is then interesting to ask what is the minimal subset that can provide sufficient location information, among a set of sensors that potentially can be queried.

In the next section, we introduce the more general Bayesian framework, within which issues such as information optimality can be formulated.

2.2.3   Bayesian State Estimation

The goal of localization or tracking is to obtain a good estimate of the target state $\mathbf{x}^{(t)}$ from the measurement history $\overline{\mathbf{z}^{(t)}}$. For this problem, we adopt a classic Bayesian formulation.

In statistical terms, we use $p(\mathbf{x})$ to denote the a priori probability distribution function (PDF) about the state $\mathbf{x}$, $p(\mathbf{z}|\mathbf{x})$ the likelihood function of $\mathbf{z}$ given $\mathbf{x}$, and $p(\mathbf{x}|\mathbf{z})$ the a posteriori distribution of $\mathbf{x}$ given the measurement $\mathbf{z}$. One can think of the likelihood function as describing the probabilistic information contained in measurement **z** about **x**; this can be represented by a sufficient statistic of **x** in practice. We also informally refer to the a posteriori distribution $p(\mathbf{x}|\mathbf{z})$ as the current *belief*.

The relationship between the a posteriori distribution $p(\mathbf{x}|\mathbf{z})$, the a priori distribution $p(\mathbf{x})$, and the likelihood function $p(\mathbf{z}|\mathbf{x})$ is given by Bayes Theorem,

$$p(\mathbf{x}|\mathbf{z}) = \frac{p(\mathbf{z}|\mathbf{x})\,p(\mathbf{x})}{\int p(\mathbf{z}|\mathbf{x})\,p(\mathbf{x})\,d\mathbf{x}} = \frac{p(\mathbf{z}|\mathbf{x})\,p(\mathbf{x})}{p(\mathbf{z})}, \tag{2.5}$$

where $p(\mathbf{z})$ is the marginal distribution, also called the *normalizing constant*. Therefore, we only need to compute the numerator, $p(\mathbf{z}|\mathbf{x})p(\mathbf{x})$, and normalize it as necessary. For notational convenience, the Bayes rule (2.5) will be written as

$$p(\mathbf{x}|\mathbf{z}) = k\,p(\mathbf{z}|\mathbf{x})\,p(\mathbf{x})$$

or

$$p(\mathbf{x}|\mathbf{z}) \propto p(\mathbf{z}|\mathbf{x})\,p(\mathbf{x}). \tag{2.6}$$

We would like our estimate $\hat{\mathbf{x}}^{(t)}$ to be, on the average, as close to the true value $\mathbf{x}^{(t)}$ as possible according to some measure. A commonly used estimator in the standard estimation theory is the so-called *minimum-mean-squared error* (MMSE) estimator. Given a set of distributed measurements, $\mathbf{z}_1, \ldots, \mathbf{z}_K$, the MMSE estimate is the expected value, or *mean*, of the distribution $p(\mathbf{x}\,|\,\mathbf{z}_1, \ldots, \mathbf{z}_N)$, which can be written as

$$\bar{\mathbf{x}} = \int \mathbf{x}\,p(\mathbf{x}\,|\,\mathbf{z}_1, \ldots, \mathbf{z}_N)\,d\mathbf{x},$$

and the residual uncertainty of the estimate is approximated by the *covariance*

$$\Sigma = \int (\mathbf{x} - \bar{\mathbf{x}})(\mathbf{x} - \bar{\mathbf{x}})^T p(\mathbf{x} \mid \mathbf{z}_1, \ldots, \mathbf{z}_N) \, d\mathbf{x}.$$

Appendix A at the end of the book describes other forms of optimal estimators.

What the standard estimation theory does not consider but which is of great importance to distributed sensor networks, is the fact that knowledge of the measurement value $\mathbf{z}_i$ and sensor characteristics λ_i normally resides only in sensor i. In order to compute the belief based on measurements from several sensors, we must pay a cost for communicating that piece of information. Thus, determining what information each sensor node needs to receive from other sensor nodes is an important decision. This is why the sensor characteristics $\lambda_i^{(t)}$ are explicitly represented, because it is important to know the type of information that each sensor might provide for various information processing tasks.

Since updating a belief with new measurements is now assigned a cost, the problem is to intelligently choose a subset of sensor measurements that provides "good" information for constructing a belief state as well as minimizing the cost of having to communicate sensor measurements to the processing node. In order to choose sensors that provide informative updates to the belief state, it is necessary to introduce a measure of the information a sensor measurement can provide to that state. In Chapter 5, we formalize criteria for sensor selection.

Centralized Estimation

Consider a centralized Bayesian estimator for a sensor network consisting of K sensors. At any time instant t, each sensor i ($i = 1, 2, \ldots, K$) informs a central processing unit about its measurement $\mathbf{z}_i^{(t)}$. The central processing unit updates the belief state using Bayesian estimation as in (2.6), with the difference that instead of using the single sensor measurement $\mathbf{z}_i^{(t)}$, it uses the measurement collection $\underline{\mathbf{z}}^{(t)} = \left\{ \mathbf{z}_1^{(t)}, \mathbf{z}_2^{(t)}, \ldots, \mathbf{z}_K^{(t)} \right\}$. If the sensor measurements are mutually independent when conditioned on the target locations,[1] then

$$p\left(\underline{\mathbf{z}}^{(t)}\middle|\mathbf{x}^{(t)}\right) = \prod_{i=1,\ldots,K} p\left(\mathbf{z}_i^{(t)}\middle|\mathbf{x}^{(t)}\right). \tag{2.7}$$

The centralized tracking algorithm utilizes all K measurements at every time step. If the communication between the sensors and the central unit is through radio, the power needed to communicate reliably is proportional to the communication distance raised to a constant power α, where $\alpha > 2$. Thus, from an energy point of view, the centralized algorithm is inefficient in utilizing the communication resources. From the processing point of view, the complexity of the centralized algorithm scales linearly with K and hence is prohibitive for large networks—not to mention being vulnerable to a single point of failure at the central unit and thus lacking in robustness.

Sequential (or Incremental) Estimation

In practice, the measurements $\mathbf{z}$ may be acquired over time, either by a single sensor or multiple sensors, and are denoted by $\overline{\mathbf{z}^{(t)}}$. We extend the basic Bayesian estimation (2.6) so that it can incrementally combine measurements over time. Suppose a sensor node has a prior belief state given previous measurements $p\left(\mathbf{x}^{(t)}\middle|\overline{\mathbf{z}^{(t-1)}}\right)$, and at time t it takes a new measurement $\mathbf{z}^{(t)}$. Assuming the following conditional independence assumptions are satisfied,

- conditioned on $\mathbf{x}^{(t)}$, the new measurement $\mathbf{z}^{(t)}$ is independent of the past measurement history $\overline{\mathbf{z}^{(t-1)}}$;

- conditioned on $\mathbf{x}^{(t-1)}$, the new position $\mathbf{x}^{(t)}$ is independent of $\overline{\mathbf{z}^{(t-1)}}$.

---

1 The conditional independence states that given knowledge about the target state **x** and assuming the sensors are well separated, the statistical perturbations in the measurements about **x** come from noise effects during signal propagation which are not correlated.

These assumptions are often satisfied in practice; we discuss exceptions when they do arise in subsequent chapters. Under these assumptions, based on the new measurement, the sensor node computes the new belief $p\left(\mathbf{x}^{(t)}|\mathbf{z}^{(t)}\right)$ using the Bayes rule:

$$p\left(\mathbf{x}^{(t)}\Big|\overline{\mathbf{z}^{(t)}}\right) \propto p\left(\mathbf{z}^{(t)}\Big|\mathbf{x}^{(t)}\right) p\left(\mathbf{x}^{(t)}\Big|\overline{\mathbf{z}^{(t-1)}}\right)$$

$$= p\left(\mathbf{z}^{(t)}\Big|\mathbf{x}^{(t)}\right) \int p\left(\mathbf{x}^{(t)}\Big|\mathbf{x}^{(t-1)}\right) p\left(\mathbf{x}^{(t-1)}\Big|\overline{\mathbf{z}^{(t-1)}}\right) d\mathbf{x}^{(t-1)}.$$

$$(2.8)$$

Note that the a posteriori distribution is recursively computed after each measurement $\mathbf{z}^{(t)}$ is made—hence the name sequential *Bayesian estimation.*

In (2.8), $p\left(\mathbf{x}^{(t-1)}\Big|\overline{\mathbf{z}^{(t-1)}}\right)$ is the belief inherited from the previous step; $p\left(\mathbf{z}^{(t)}\Big|\mathbf{x}^{(t)}\right)$ is the likelihood of observation given the target state. The function $p\left(\mathbf{x}^{(t)}\Big|\mathbf{x}^{(t-1)}\right)$ encodes target dynamics. For example, if a vehicle target is moving at a known velocity $\mathbf{v}$, then $p\left(\mathbf{x}^{(t)}\Big|\mathbf{x}^{(t-1)}\right)$ is simply $\delta(\mathbf{x}^{(t)} - \mathbf{x}^{(t-1)} - \mathbf{v})$, where $\delta(\cdot)$ is the Dirac delta function. In practice, however, the exact vehicle velocity is rarely known. For example, we may assume that the vehicle has a speed (i.e., distance traveled per sample interval) uniformly distributed in $[0, v_{max}]$, and the vehicle heading is uniform in $[0, 2\pi)$. In this case, $p\left(\mathbf{x}^{(t)}\Big|\mathbf{x}^{(t-1)}\right)$ is a disk centered at $\mathbf{x}^{(t-1)}$ with radius v_{max}. Under this model, the predicted belief $p\left(\mathbf{x}^{(t)}\Big|\overline{\mathbf{z}^{(t-1)}}\right)$ [i.e., the integral in (2.8)] is obtained by convolving the old belief $p\left(\mathbf{x}^{(t-1)}\Big|\overline{\mathbf{z}^{(t-1)}}\right)$ with the uniform circular disk kernel. The convolution reflects the dilated uncertainty about target location due to uncertain motion. Figure 2.3 shows an example of the various distributions and likelihood function in a Bayesian estimation. Note that they are all functions of the state $\mathbf{x}$, in this case, the position in a two-dimensional plane.

The Kalman filter [110] is a special case of (2.8), where the belief distributions and error models are Gaussians and the system dynamics model is linear. For many vehicle tracking problems, the

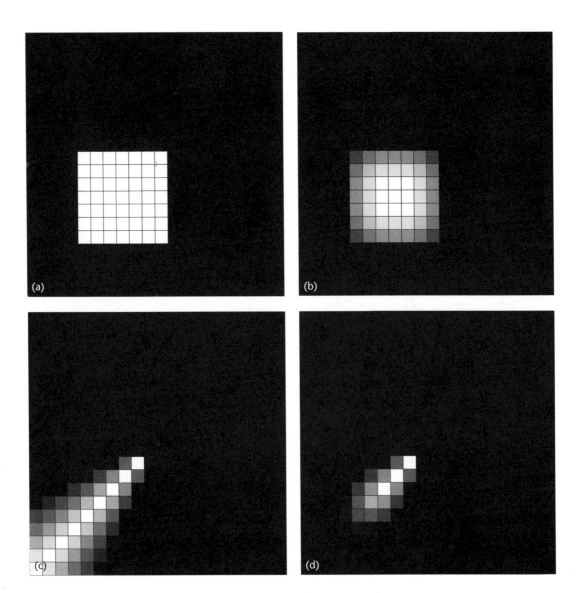

Figure 2.3 Computation of a posteriori distribution about a target position using sequential Bayesian estimation. (a) A posteriori distribution $p\left(\mathbf{x}^{(t-1)}\middle|\mathbf{z}^{(t-1)}\right)$ from time $t-1$. (b) The predicted belief $p\left(\mathbf{x}^{(t)}\middle|\mathbf{z}^{(t-1)}\right)$ is obtained by convolving the a posteriori distribution with the target dynamics $p\left(\mathbf{x}^{(t)}\middle|\mathbf{x}^{(t-1)}\right)$. (c) $p\left(\mathbf{z}^{(t)}\middle|\mathbf{x}^{(t)}\right)$ is the likelihood of observation at time t. (d) The a posteriori distribution $p\left(\mathbf{x}^{(t)}\middle|\mathbf{z}^{(t)}\right)$ at time t is computed by combining the likelihood function and predicted belief.

sensor models may be multimodal and non-Gaussian, and the vehicle dynamics may be highly nonlinear. Hence, we will use the more general Bayesian formulation of the tracking problem (2.8) when we develop distributed tracking techniques.

2.3 Distributed Representation and Inference of States

How do we represent the state information of a target so that efficient computation can be performed to infer the target state? To address this, we need to consider both the representation of state information, or model, as well as the mapping of the representation to distributed sensors.

2.3.1 Impact of Choice of Representation

There are several ways to approximate an arbitrary belief state regarding the targets:

1. We can approximate the belief by a family of distributions $M \subset \mathcal{P}(\mathcal{R}^d)$, parameterizable by a finite dimensional vector parameter $\theta \in \Theta$, where $\mathcal{P}(\mathcal{R}^d)$ is the set of all probability distributions over $\mathcal{R}^d$. An example of M is the family of Gaussian distributions on $\mathcal{R}^d$ where the finite dimensional parameter space Θ includes the mean and covariance of the Gaussian distributions. In these cases, many efficient prediction and estimation methods exist, such as the Kalman filter.

2. We can approximate the belief by weighted point samples. This is a brute force method of discretizing the probability distribution (or density) of a continuous random variable by a probability mass function (PMF) with support in a finite number of points of S. Let $\tilde{S} \subset S$ be some finite set of points of S. Then, we can approximate the density by a PMF with mass at each point of $\tilde{S}$ proportional to the density at that point. Two examples of this approximation are (1) discretizing the subset of S by a grid and (2) the particle filter approximation of distributions. A variant of this point sample

approximation is to partition S and assign a probability value to each region—a histogram-type approach.

We refer to the first approximation as *parametric* and the second approximation as *nonparametric*. Although the terms parametric and nonparametric may seem to imply two completely different ways of approximation, it is worth pointing out that the distinction is not clear cut. In the second case, if we consider the points of $\tilde{S}$ with their associated probability value as the set of all parameters Θ, then we can consider the second approximation a parametric approach as well. However, the approximation in the second case cannot be characterized by a *small* number of parameters.

The representation of the belief impacts the amount of communication resource consumed when sending the belief state to a remote sensor. If we must pass the belief to a remote sensor, we are faced with the following trade-off:

- Representing the true belief by a parametric approximation with relatively few parameters will result in a poor approximation of the belief state, but with the benefit that fewer bits need to be transmitted. On the other hand, representing the true belief by a nonparametric approximation will result in a more accurate approximation of the belief at the cost of more transmitted bits.

The above trade-off is a general statement. However, there is another factor underlying such trade-offs. That factor is background knowledge of the possible set of beliefs. Let us elaborate.

Parametric Representation (e.g., Gaussian Approximation)

The reason we are able to transmit relatively few bits in the parametric case is due to the assumption that all sensors are aware of the parametric class of beliefs. Knowledge of this parametric class is the background knowledge that allows for only a small number of bits to be transmitted. In the case where we approximate the belief by a Gaussian, all sensors know that the belief is represented by the parametric class of Gaussian distributions. Hence, only the mean and covariance need to be transmitted. The Kalman filter equations are recursive update equations of the mean and covariance of the Gaussian distribution.

Measurement-Based Representation

For the nonparametric case, there is no constant-size parameterization of the belief in general. However, assuming that the model of the measurements is known, we can parameterize the belief by storing a history of all measurements. Without loss of generality, assume each measurement is a scalar z, that is, $z \in \mathcal{R}$. In this case, the parameter space $\Theta = \mathcal{R}^*$ is the set of all finite length strings of real numbers. If $\theta^m = z^1 z^2 \cdots z^m$ is the parameter for the likelihood function

$$p(z^1, \ldots, z^m \mid \mathbf{x}),$$

then the parameter $\theta^{m+1} = z^1 z^2 \cdots z^m z^{m+1}$ for

$$p(z^1, \ldots, z^{m+1} \mid \mathbf{x})$$

is updated from θ^m by

$$\theta^{m+1} = \theta^m \cdot z^{m+1},$$

where "$\cdot$" denotes string concatenation. This update equation is trivial and suffers from increasing dimensionality. But if we are given knowledge of a highly nonlinear model for the measurements of the unknown variable we wish to estimate, it may be that collecting a history of the measurements is a more compact representation initially. To elaborate, a Gaussian approximation of the likelihood function for two independent measurements z_1 and z_2 of a two-dimensional target state $p(z_1, z_2 \mid \mathbf{x})$ would require storing the mean and covariance, which for this particular case would require two real numbers for the mean and three real numbers for the covariance, for a total of five real numbers. However, to store the exact likelihood, we require only two real numbers corresponding to z_1 and z_2, with the implicit background knowledge of the form of the likelihood function.

After some number of measurements, it may turn out that the true likelihood function becomes unimodal and, hence, can be approximated well by a Gaussian. This motivates a *hybrid* representation:

1. Initially, the belief is parameterized by a history of measurements.

2. Once the belief begins to look unimodal, we will approximate the belief by a Gaussian and, from here on, the parameterization of the belief includes the mean and covariance.

Since a Gaussian approximation is poor initially, the cost of maintaining a history of measurements is justified. Once the Gaussian approximation becomes reasonable, we can convert to such a representation and reduce the parameter dimensionality.

Discrete Samples

As we shall see in Section 2.5, the observation models for acoustic amplitude and DOA sensors are nonlinear. Consequently, the likelihood $p\left(\mathbf{z}^{(t)}\middle|\mathbf{x}^{(t)}\right)$ is non-Gaussian, as is the a posteriori belief $p\left(\mathbf{x}^{(t)}\middle|\mathbf{z}^{(t)}\right)$. For these non-Gaussian distributions, one may use a grid-based nonparametric representation for probability distributions. The distributions are represented as discrete grids in a d-dimensional space. An example of the belief plots in a two-dimensional plane is given in Figure 2.3. In that figure, the gray level depicts the probability distribution function (PDF) evaluated at the grid location. The lighter the grid square, the higher the PDF value.

The grid representation of likelihood and posterior belief admits an efficient computation. The MMSE estimate is simply the average of the grid locations in the belief cloud weighted by the belief value. The predicted belief $p\left(\mathbf{x}^{(t)}\middle|\mathbf{z}^{(t-1)}\right)$ [the integral in (2.8)] is computed by a convolution of the original belief cloud with the target dynamics [see, e.g., Figure 2.3(b)].

The resolution of the grid representation is constrained by the computational capacity, storage space, and communication bandwidth of the sensor nodes. For the acoustic sensors, the likelihood functions are relatively smooth. This smoothness permits the use of low-resolution representation without much loss in performance. Furthermore, a sparse representation can be used that only stores the grid points whose likelihood values are above a prespecified threshold.

An alternate representation is to use a set of samples that are not necessarily on a regular grid. This is the basis for the so-called particle filter, which is described in detail in Appendix B at the end of the book.

2.3.2 Design Desiderata in Distributed Tracking

A target tracker using the Bayesian filter (2.8) may be implemented in a distributed sensor network in a number of ways. The belief state $p\left(\mathbf{x}^{(t)}\middle|\overline{\mathbf{z}^{(t)}}\right)$ can be stored centrally at a fixed node, at a sequence of nodes through successive hand-offs, or at a subset of nodes concurrently. In the first case [Figure 2.4(a)], a fixed node is designated to receive the measurements from other relevant sensors through communication. This simpler tracker design is obtained at the cost of potentially excessive communication and reduced robustness to node failure. When a target is stationary and in the neighborhood of the designated node, the communication cost may be manageable, since the node only needs to retrieve data from sensors near the target. In the more common moving-target scenario, the communication cost could become prohibitive.

In the second case [Figure 2.4(b)], an initial leader node is elected and that node combines measurements from (hopefully) nearby sensors. As the target moves in the sensor field, task requirements change, or other environmental conditions vary, the initial leader is no longer desirable. A new leader is elected to collect data from its nearby sensors. The current leader hands off the current belief to the new leader, who then updates the belief with its new data. Because of continuity in target motion, it is rare that the new leader will be many communication hops away from the current leader node. The leader election and data hand-off process is iterated until the task objective is satisfied. This scheme of state representation and storage localizes the communication to those sensors in a geographical neighborhood

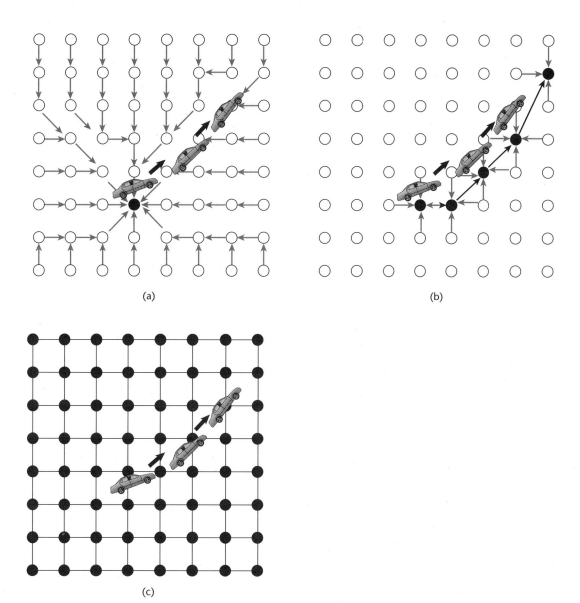

Figure 2.4 Storage and communication of target state information in a sensor network. In the figures, circles on the grid represent sensor nodes, and some of the nodes (i.e., solid circles) are those that store target state information. Narrow, gray arrows or lines denote communication paths among the neighbor nodes. Narrow, black arrows denote sensor hand-offs. A vehicle target moves through the sensor field, as indicated by the thick arrows. (a) A fixed node stores the target state. (b) Leader nodes are selected in succession according to information such as vehicle movement. (c) Every node in the network stores and updates target state information (adapted from [232]).

of the targets and can be advantageous in reducing the overall bandwidth consumption and increasing the lifetime of the network. One of the key research challenges here is to define an effective selection criterion for sensor leaders that can be easily evaluated by each sensor; this is the main topic for Chapter 5. In the single moving leader design, the robustness of the tracker may suffer from occasional leader node attrition. One way to mitigate the problem is to replicate the belief state in nearby nodes in a consistent manner and to detect and respond to leader failure. The moving leader design can also be used for tracking multiple targets in a sensor field, with one leader node tracking each target. Special care must be taken when targets move into proximity of each other.

Finally, the belief state can be stored in a distributed fashion across a section of sensor nodes. This is attractive from the robustness point of view. The sensing and estimation can all be accomplished nodewise, thus localizing the communication. The major design challenge is to efficiently infer global properties about targets, some of which may be discrete and abstract, from partial and local information, and to maintain information consistency across multiple nodes. Consider the multitarget case, where the local state information might be stored in possibly overlapping sections of sensor nodes in the network. The distributed target state information has to be organized so that a new piece of data can be easily combined with the appropriate target state at a sensor node. Moreover, the state information needs to be organized to allow user queries to discover, retrieve, and aggregate the information efficiently. Many issues about the leaderless distributed tracker are still open and deserve much attention from the research community.

As we have seen, information representation, storage, and access are the key problems in designing distributed CSIP applications. Solving these under the resource constraints of a sensor network may require new routing and aggregation techniques that differ considerably from the existing TCP/IP end-to-end communication. Chapter 3 thus describes networking techniques for sensor networks. The research community has been searching for the right "sensor net stack" that can provide suitable abstractions over networking and hardware resources. While defining a unifying architecture for sensor networks is still an open problem, a key element of such an architecture is the *principled interaction between the application and networking layers*. Chapter 5 describes a number of approaches that express application requirements as a set of information and cost constraints, so that an ad hoc networking layer can effectively support the application.

2.4 Tracking Multiple Objects

We have considered the estimation problem for tracking a single target by a sensor network. Tracking multiple interacting targets distributed over a geographical region is significantly more challenging for two reasons:

1. *Curse of dimensionality*: The presence and interaction of multiple phenomena cause the dimension of the underlying state spaces to increase. Recall that the joint state space of multiple targets is a product space of individual state spaces for the targets. Estimating the phenomenon states jointly suffers from the state-space explosion, since the amount of data required increases exponentially with the dimension. This is inherent in any high-dimensional estimation problem, regardless of whether the sensing system is centralized or distributed.

2. *Mapping to distributed platforms*: An estimation algorithm for tracking multiple targets will have to be mapped to a set of distributed sensors, as will the state-space model for the estimation problem. To ensure the responsiveness and scalability of the system, the communication and computation should be localized to relevant sensors only.

2.4.1 State-Space Decomposition

The first challenge noted earlier is addressed by developing a distributed state-space representation that systematically decomposes a high-dimensional estimation problem into more manageable lower-dimensional subproblems. The basic idea is to factor a joint state space of all targets of interest, whenever possible, into lower-dimensional subspaces, each of which can be represented and processed in a distributed fashion. Efficient approximate algorithms for maintaining consistency across the subspaces will have to be developed. The second challenge is addressed by developing a *collaborative group abstraction* in which the state-space model is mapped onto a sensor network via dynamically formed collaborative groups. Group structures must be designed to create and maintain the necessary state models in suitable joint or factored spaces and to support communications among them. Collaborative groups can effectively localize the computation and communication to sensors who must share and manipulate the common state information. The collaborative group management is a subject of study in Chapters 5 and 7.

The key idea in the state-space factorization is to decouple information in a state space into location and identity information. This allows the tracking problem to be solved separately by location estimation, which requires frequent local communication, and by identity management, which requires less frequent, longer range communication. The location tracking problem can be carried out in the subspaces, or marginal spaces, when targets are far apart. When two targets move closer together, tracking them separately will incur increasingly larger errors because of target interaction and signal mixing. Hence, they have to be estimated jointly. Figure 2.5 illustrates the transitions from factored to joint spaces and then back to factored spaces. The decision of when to switch to another representation depends on several factors such as localization accuracy.

However, the target identity information may be lost when we switch from a joint space representation to factored spaces. This can be illustrated in the following simple example. Assume the state space for each target is one-dimensional, and the joint belief $p(x_1, x_2)$ is symmetric in x_1 and x_2. Figure 2.6(a) shows such an example, where there are two probability mass "blobs" centered at (a, b) and (b, a) in a

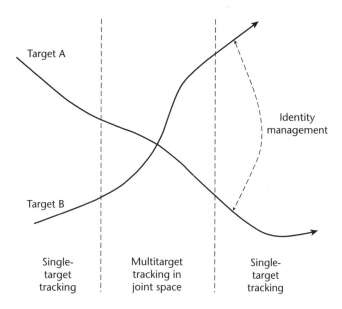

Figure 2.5 Modes of interaction in a multitarget tracking problem. The corresponding state representations move from factored to joint and then back to factored spaces (adapted from [146]).

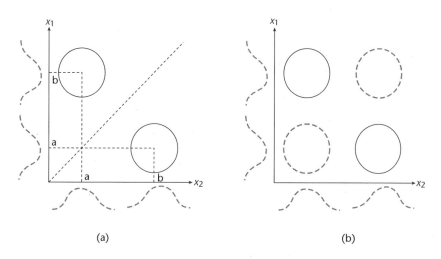

Figure 2.6 Switching between joint density $p(x_1, x_2)$ and marginal densities $p(x_1)$ and $p(x_2)$ (adapted from [146]).

x_1-x_2 joint state space. Let the factorization be done by the following marginalization:

$$p(x_1) = \int p(x_1, x_2)\, dx_2, \quad p(x_2) = \int p(x_1, x_2)\, dx_1.$$

This gives the marginal distributions shown as dotted curves along the axes of Figure 2.6(a). The distributions are multimodal. Each target has an equal probability of being at a or b. Now we reconstruct the joint space by taking a product of the marginals:

$$p(x_1, x_2) = p(x_1)p(x_2).$$

The reconstructed joint space now has four "blobs" [Figure 2.6(b)]. But we know that two of them are not feasible since the targets cannot be both at a or both at b. This illustrates that the target identity information has been lost from the joint space to the factored space. We will need to maintain the target identity information properly, using, for example, an identity management algorithm [210].

2.4.2   Data Association

Another approach to addressing the multitarget tracking problem is to use data association techniques to attribute each sensor measurement to a target signal source. In this approach, the target location and identity information are coupled and processed in the same space. When multiple targets are present, one has to associate a measurement with a correct target before applying an estimation technique such as the Bayesian method on the measurements for that target [13]. An association hypothesis assigns a set of measurements to a set of targets, and the total number of possible associations is exponential in the number of sensors.

As analyzed earlier, the state space $\mathcal{X}$ of N targets can be understood as a product of individual target state spaces $\mathcal{X}_n$; that is, the state vector is $\mathbf{x} = [\mathbf{x}_1, \ldots, \mathbf{x}_N]^T \in \mathcal{X}$. For a network of K sensors, the sensor output vector is $\mathbf{z} = [\mathbf{z}_1, \ldots, \mathbf{z}_K]^T$, where $\mathbf{z}_i$ is the output of sensor i. Each $\mathbf{z}_i$ could be a measurement of a single target or multiple targets.

The problem of sequential state estimation in a multitarget, multisensor environment can be stated as follows. At each time step t, given the previous state estimate at $t-1$ and the current observation, the general estimation problem can be stated as a mapping:

$$\mathcal{X}^{(t-1)} \times \mathcal{Z}^{(t)} \to \mathcal{X}^{(t)}. \tag{2.9}$$

One can think of the Bayesian estimation (2.8) as a special case.

To estimate $\mathbf{x}^{(t)} \in \mathcal{X}^{(t)}$, an observation $\mathbf{z}_i$ contributed by a target j must be associated with the correct target state $\mathbf{x}_j$ in order to carry out the estimation (2.9). We illustrate the computational difficulties of data association for the estimation problem for two cases.

- **Case I**: Assume there is no signal mixing and each observation $\mathbf{z}_i$ measures a signal $\mathbf{g}_n \in \mathcal{G}$ from target n only. When each observation corresponds to a unique target, then $\{1, \ldots, K\}$ is a permutation of a K-subset of the set $\{1, \ldots, N\}$. The number of possible associations of $\mathbf{z}_i$'s with the corresponding $\mathbf{x}_j$'s is $C_N^K = \frac{N!}{K!}$. If multiple observations may come from a single target, then the number of associations is N^K. In either case, the complexity is exponential in the number of measurements at each time step.

- **Case II**: More generally, each sensor observation $\mathbf{z}_i$ measures a composite of $\mathbf{g}_n$'s through a mixing function: $H : \mathcal{G}^{(t)} \to \mathcal{Z}^{(t)}$. Without prior knowledge about H, any combination of $\mathbf{g}_n$'s could be present in the $\mathbf{z}_i$'s. Pairing each $\mathbf{z}_i$ with each subset of the $\mathbf{g}_n$'s creates 2^N associations. The total number of associations of $\mathbf{z}$ with $\mathbf{x}$ is then 2^{NK}, that is, exponential in both the numbers of sensors and targets (signal sources).

For applications such as tracking, it may be necessary in some cases to reason across multiple time steps and examine the history of all possible state evolutions (or tracks)—for example, in order to roll back states. Each pairing of observations with states in the above single-step state estimation creates a hypothesis of the state evolution. As more observations are made over time, the total number of possible state evolution sequences is exponential in the numbers of sensors *and* measurements over time.

To address the data association problem, a number of approaches have been developed by the multisensor multitarget tracking research community. These include Multiple Hypothesis Tracking (MHT) [193] and Joint Probabilistic Data Association (JPDA) [14, 190]. MHT and JPDA were specifically designed for managing association hypotheses in multiple target tracking. MHT forms and maintains multiple association hypotheses. For each hypothesis, it computes the probability that it is correct. On the other hand, JPDA evaluates the association probabilities and combines them to compute the state estimate. Straightforward applications of MHT and JPDA suffer from the combinatorial explosion in possible data associations. Knowledge about targets, environment, and sensors can be exploited to rank and prune hypotheses [48, 49, 183].

These methods, however, were primarily developed for centralized tracking platforms. Relatively little consideration was given to the fundamental problems of moving data across distributed sensor nodes in order to combine the data and update track information. There was no cost model for communication in the tracker. Furthermore, sensor data held up by communication delays may not arrive at a tracking node in the same sequence recorded by the time stamps with which the data was generated. Yet the Kalman or Bayesian filter assumes a strict temporal order on the data during the sequential update, and so may have to roll back the tracker in order to incorporate a "past" measurement or may decide to throw away the data.

The distributed Kalman filter, as described in [152], is a global method, requiring each sensor node to communicate its measurement to a central node, where estimation and tracking are carried out. In this method, sensing is distributed while tracking is centralized. A significant amount of communication overhead is incurred in collecting the data at the central node. Moreover, this approach relies on a communication network to link the central node with the rest of the sensors in the network, rendering it more susceptible to link or node failure. In the case where the sensor measurements from multiple nodes are not independent (when the Kalman filter is not applicable), a covariance intersection algorithm may be used to combine the data [218, 40]. More recently, other distributed approaches that exploit peer-to-peer networking in sensor networks have been developed [232, 41].

Replicated information is another serious problem in distributed tracking, whether the tracking is about a single target or multiple targets. One source of information double counting is due to loopy propagation of evidence in a network. A single piece of evidence may be used in a Bayesian estimation multiple times, resulting in an overly confident estimate. This is similar to the loopy problem in Bayesian networks [177] and may be addressed by approximation algorithms developed in the Bayesian network research community. Another source of information double counting is due to multiple sensor nodes observing a single target and reporting multiple detections. An aggregation module will be necessary to combine and consolidate multiple local detections before reporting the overall detection to a user. Since network delays can cause local detections to arrive at different, unpredictable times, the aggregation module may need to compare detections over an extended time period, assuming the detections are all time stamped.

2.5 Sensor Models

We describe two common types of sensors for tracking: acoustic amplitude sensors and direction-of-arrival (DOA) sensors. An acoustic amplitude sensor node measures sound amplitude at the microphone and estimates the distance to the target based on the physics of sound attenuation. An acoustic DOA sensor is a small microphone array. Using beam-forming techniques, a DOA sensor can determine the direction from which the sound comes, that is, the bearing of the target. More generally, range sensors estimate distance based on received signal strength or time difference of arrival (TDOA), while DOA sensors estimate signal bearing based on TDOA.

The nonparametric representation we described in Section 2.3 poses few restrictions on the type of sensors to be used and allows the network to easily fuse data from multiple types of sensors. Relatively low-cost sensors such as microphones are attractive because of their affordability as well as simplicity in computation. However, there are no barriers to adding other sensor types, including imaging, motion, infrared (IR), or magnetic sensors. Low-cost, integrated imagers, or cameras, are attractive for target detection and identification because of the richness in the information an image can capture. Video cameras provide additional motion information that can be very useful in recognizing subtle behaviors of subjects. For example, localization or tracking of people in a complex environment using multiple cameras is described in [30, 45, 162]. A sensor network can utilize a variety of sensing modalities to optimally sense phenomena of interest.

For sensor network applications, a sensor may be characterized by properties such as cost, size, sensitivity, resolution, response time, energy usage, and ease of calibration and installation. One may have to carefully balance the utility of a sensor with the cost of processing the data. A major consideration is whether the data processing involves only local data or requires data from a number of sensors. For example, to estimate the bearing of a target, one needs to combine measurements from several sensors through communication if the sensors are not on a single node. This adds to the cost of processing and places an additional requirement on time synchronization across nodes. To ground the discussions, we will study the acoustic amplitude and DOA sensors.

Acoustic Amplitude Sensor

Assuming that the sound source is a point source and sound propagation is lossless and isotropic, a root-mean-squared (RMS) amplitude measurement z (a scalar in this case) is related to the sound source position $\mathbf{x}$ as

$$z = \frac{a}{\|\mathbf{x} - \zeta\|} + w, \tag{2.10}$$

where a is the RMS amplitude of the sound source, ζ is the location of the sensor, and w is RMS measurement noise [115]. This is a special case of equation (2.4). For simplicity, we model w as a Gaussian with zero mean and variance σ^2. The sound source amplitude a is also modeled as a random quantity. Assuming a is uniformly distributed in the interval $[a_{lo}, a_{hi}]$, the likelihood has a closed-form expression:

$$
\begin{aligned}
p(z|\mathbf{x}) &= \int_{a_{lo}}^{a_{hi}} p(z|\mathbf{x}, a) p(a) \, da \\
&= \frac{1}{\Delta a} \int_{a_{lo}}^{a_{hi}} \frac{1}{\sqrt{2\pi\sigma^2}} e^{-\frac{(z-\frac{a}{r})^2}{2\sigma^2}} \, da \\
&= \frac{r}{\Delta a} \left[\Phi\left(\frac{a_{hi} - rz}{r\sigma}\right) - \Phi\left(\frac{a_{lo} - rz}{r\sigma}\right) \right],
\end{aligned}
\tag{2.11}
$$

where $\Delta a = a_{hi} - a_{lo}$, $r = \|x - \zeta\|$ is the distance between the sound source and the sensor, and $\Phi(\cdot)$ is the standard error function. Interested readers can find details of the derivation in [43].

Figure 2.7(a) shows an example of the likelihood function $p(z|\mathbf{x})$, a doughnut-shaped function centered at the sensor location. The thickness of the annulus (outer radius minus inner radius) is determined by a_{lo}, a_{hi}, and σ^2. Fixing the first two, the thickness increases with σ^2, as the target location is more uncertain. Fixing σ^2, the thickness increases as a_{lo} decreases or as a_{hi} increases. In Cartesian space, it is clear that this likelihood is non-Gaussian and difficult to approximate as a sum of Gaussians. The cross section of the likelihood function along the radial direction is plotted in Figure 2.7(b), and it is quite smooth and amenable to approximation by sampling.

The uniform, stationary assumption of the source amplitude gives rise to efficient computation in estimation. To accommodate quiet targets—for example, quiet vehicles or vehicles in an idle state—a_{lo}

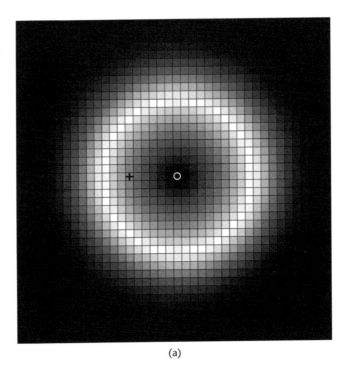

(a)

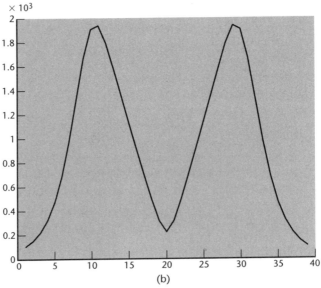

(b)

Figure 2.7   Likelihood function $p(z|\mathbf{x})$ for acoustic amplitude sensors. (a) A grid representation. The circle at the center denotes the sensor location ζ, and the small cross the true target location. (b) The cross section of the likelihood along the horizontal line past the sensor location (adapted from [144]).

can be set to zero in practice, and a_{hi} can be set via calibration. This uniform assumption is simplistic and may be replaced by better models of source amplitude that exploit the correlation of sound energy over time, but at a much higher computational cost.

DOA Sensor

Amplitude sensing provides a range estimate. This estimate is often not very compact (i.e., not unimodal) and is limited in accuracy due to the crude uniform source amplitude model. These limitations make the addition of a target-bearing estimator very attractive.

Beam-forming algorithms are commonly used in radar, speech processing, and wireless communications to enhance signals received at an array of sensors. Assume that we have a microphone array composed of M identical omnidirectional microphones. The data collected at the mth microphone at time t is

$$g_m(t) = s_0(t - t_m) + w_m(t), \tag{2.12}$$

where s_0 is the source signal, w_m is the noise (assumed to be white Gaussian), and t_m is the time delay, which is a function of the direction of arrival θ. In a far-field problem, the sound source is sufficiently far away from the microphone array so that the wave received at the array can be considered a planar wave (Figure 2.8). The relative time delay between two sensors with a spacing d can be expressed as a function of the bearing angle θ,

$$t_m = \frac{d}{c} \sin \theta,$$

where c is the speed of sound propagation. A simple delay-and-sum of the signals at the sensors can coherently enhance the signal. Thus, a conventional beam-former performs a kind of spatial filtering. For the sensor network context, we are also interested in estimating the signal bearing angle θ.

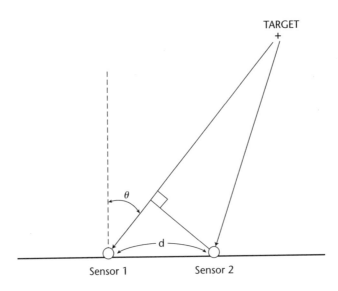

Figure 2.8 Coherent spatial signal processing using a sensor array. The signal impinging on the sensors is a planar wave under the far-field assumption.

A number of DOA estimation algorithms have been developed. These methods include the least-squares and maximum likelihood (ML) methods.[2] The least-squares method solves for target bearing and/or location by estimating first the time difference of arrival at a set of sensors and then the source location. This is particularly useful for near-field problems in which the target is relatively close to the sensors compared with the sensor spacing. Ranging using the least-squares method is described in some detail in Chapter 4. The ML method typically converts a set of signals from time domain to frequency domain and then performs a possibly multidimensional search to find the peak in the correlation spectra [35, 36]. The ML method can be applied to both near- and far-field problems. It works well for wideband acoustic signals and does not require the microphone array to be linear or uniformly spaced. However, few DOA algorithms address problems where multiple sources are present.

---

2 See Appendix A, in back of book, for a definition of the ML method.

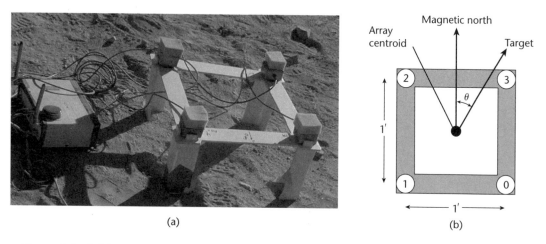

(a)   (b)

Figure 2.9 DOA sensor: (a) WINS NG 2.0 sensor node with a DOA-sensing microphone array. (b) DOA sensor arrangement and angle convention (adapted from [144]).

As a concrete example, we describe the characteristics of a DOA array with four microphones, as shown in Figure 2.9. The centroid of the array is at $\zeta_0 = (0, 0)$, and the arrival angle θ is defined as the angle from the vertical axis. Due to the presence of noise, the DOA estimate of the angle $\hat{\theta}$ is perturbed from the true angle θ. Therefore, the likelihood function $p(z|\theta)$ of the DOA measurement needs to be characterized before it can be used in a Bayesian target tracker.

We use a Gaussian model with zero mean to characterize the uncertainty in the angle measurement z. The likelihood takes the form

$$p(z|\theta) = \frac{1}{\sqrt{2\pi\sigma^2}} e^{-\frac{(z-\theta)^2}{2\sigma^2}}, \qquad (2.13)$$

where θ is determined by the geometry of the sound source position **x** and the sensor position ζ. This model ignores the periodicity of angles around 2π and is a reasonably good model when the variance σ^2 is small.

We observe that the DOA estimate is more reliable in the medium distance range and is less reliable when the sound source is too close or far away from the microphone array. This is to be expected. In the near field, the planar wave assumption is violated. In the far field, the SNR is low, and the DOA estimation may be strongly influenced by noise and fail to converge to the correct angle. To account for these factors, an empirical model for the likelihood function may be developed. The model varies the standard deviation σ in the angle estimate according to the distance to the target, as illustrated in Figure 2.10(a). In the mid-range, $[r_{near}, r_{far}]$, the DOA estimation performs quite reliably and assumes a fixed standard deviation. For the near field $[0, r_{near})$, as distance decreases, the standard deviation increases linearly to account for the increasing uncertainty of DOA readings. Likewise, in the far field $(r_{far}, +\infty)$, the standard deviation increases with the distance. Under this model, the likelihood function $p(z|\theta)$ is plotted in Figure 2.10(b). It has a cone shape in the working range and fans out in the near and far ranges. It is clear that the likelihood in the two-dimensional Cartesian plane is not compact. The sequential Bayesian filtering approach (as described in Section 2.2.3) has the flexibility to accommodate such noncompactness, while a standard Kalman filtering approach may have difficulty here.

2.6 Performance Comparison and Metrics

Since a sensor network is designed for tasks such as detection, tracking, or classification, comparison or measure of performance is only meaningful when it is discussed in the context of these tasks. Here are some of the commonly used measures of performance for these tasks:

- *Detectability*: How reliably and timely can the system detect a physical stimulus? This may be measured by sensor coverage, detection resolution, dynamic range, or response latency.

- *Accuracy*: How well does the system detect, localize, or track physical stimuli? Accuracy is typically characterized in terms of tracking errors (e.g., deviation, smoothness, continuity) or detection and classification errors (e.g., false alarms or misses).

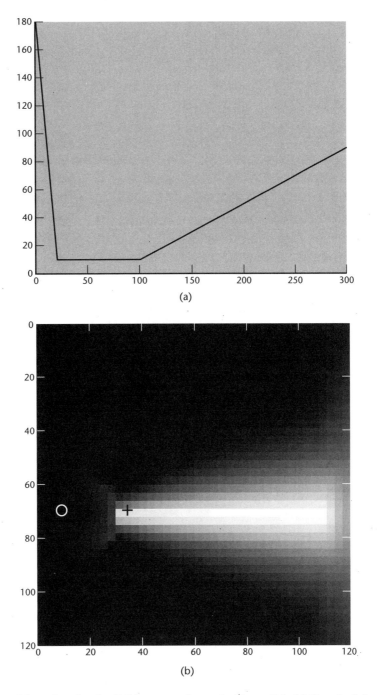

Figure 2.10 Likelihood function for the DOA sensor shown in Figure 2.9. (a) Standard deviation σ in bearing estimation versus range. Here, $r_{near} = 20$ meters, $r_{far} = 100$ meters, and $\sigma = 10°$ in $[r_{near}, r_{far}]$, based on an experimental calibration. (b) Likelihood function $p(z|\mathbf{x})$, plotted on a two-dimensional grid of the x-y space. The circle denotes the sensor location, and the cross the target location (adapted from [144]).

- *Scalability*: How does a specific property of the system vary as the size of the network, the number of physical stimuli, or the number of active queries increases?

- *Survivability*: How does the system perform in the presence of node or link failures as well as malicious attacks? Sometimes this is also called *robustness*.

- *Resource usage*: What is the amount of resources that each task consumes? The resources include energy and bandwidth.

For example, for detection and classification problems, performance is typically measured by false alarms (i.e., false positives) and misses (i.e., false negatives). A false positive occurs when the outcome is incorrectly predicted as a positive when it is in fact a negative. A false negative is when the outcome is incorrectly labeled as a negative when in fact it is a positive. To characterize the performance of a detector, it is common to use the so-called *receiver operating characteristic* (or ROC) curve, which is the detection probability versus false positive rate.

For tracking problems, performance goals and measures can be stated in terms of target and system parameters. Table 2.1 gives an example. Note that in the table, measures for different systems can only be compared for identical application scenarios. A number of measures are further explained below.

- *Source SNR*: This is measured as the SNR at a reference distance from the signal source. For an acoustic source, this is defined as the log ratio of sound pressure level (SPL) of source at the reference distance over SPL of background noise.

- *Obstacle density*: This may be measured by the probability of line-of-sight obstruction for a randomly placed target-sensor pair. It is useful in characterizing sensors such as imagers, but less useful when multipath effects on the signal are significant.

- *Network sleep efficiency*: The number of hours of target tracking versus the total number of node hours in *fully awake* mode for the entire network.

Table 2.1 Sample performance measures and parameters for tracking and localization problems.

Performance Measures	System and Application Parameters
Detection robustness (% missed and % false alarm)	Source SNR # Distractors
Detection spatial resolution (% counting error)	Intertarget spacing # Nodes
Detection latency (event occurrence to query node notification)	Link delay # Simultaneous targets # Active queries
Classification robustness (% correct)	Source SNR # Distractors
Track continuity (% track loss)	Sensor coverage area # Nodes # Simultaneous targets # Active queries Target maneuvers Obstacle density
System survivability (network partition time; % track loss)	% Node loss
Cross-node DOA estimation (bearing error)	Link capacity
Power efficiency	Active lifetime Sleep lifetime Sleep efficiency

- *Percentage of track loss*: Percentage of runs of a full scenario where continuity of vehicle identity is not correctly maintained.

2.7 Summary

This chapter has introduced the tracking problem as a representative problem for studying a number of information processing issues for sensor networks. While we have focused on the probabilistic formulation of the tracking problem using Bayesian estimation theory, the issues of how to represent states of physical phenomena, cope with the curse of the dimensionality, and map the state-space model to distributed sensors are common to many sensor network problems regardless of the choice of the formulation.

As we have discussed, the distributed representation of state information plays a key role in designing sensor network processing algorithms. Because of the decentralized nature of sensor network data collection, the a priori world model as well as the representation of the state information must reside in sensor nodes across the network in order to process the information in situ. This raises a number of interesting design issues, such as how to distribute the model and state representation across the nodes, how to update the distributed information and keep it consistent, and how to dynamically assign sensors to process different pieces of the state information. The sensor tasking problem will be dealt with in Chapter 5.

3

Networking Sensors

As the very name implies, networking is a key part of what makes sensor networks work. Networking is what allows geographical distribution of the sensor nodes and their placement close to signal sources. Effective internode communication is essential for aggregating data collected from sensor nodes in different locations, for time synchronization and node localization during network initialization, for sensor tasking and control, and so on. At the same time, radio communication is the most expensive operation a node performs in terms of energy usage, and thus it must be used sparingly and only as dictated by the task requirements. In addition, unlike their more planned and stable wired counterparts, sensor networks are typically deployed in an ad hoc manner, and unstable links, node failures, and network disconnections are all realities that have to be dealt with.

In this chapter, we discuss a number of important topics related to networking sensor nodes. These include routing algorithms, load balancing and energy awareness, and publish-and-subscribe schemes to bring information producers and consumers together. As we will see later, networking not only provides an essential layer of functionality in sensor networks, but it also integrates with application-level processing in ways that are far more synergistic than in classical wired or wireless communication networks.

3.1 Key Assumptions

The techniques developed in this chapter are based on the following assumptions about the sensor network:

- Wireless communication between nodes utilizes radio links; each node talks directly only to its immediate neighbors within radio range. Within this range, communication is by broadcast: all immediate neighbors hear what a node transmits.

Most often we will make the assumption that the radio range for a node is a disk of radius r around the node (r is taken as the same for all nodes). The connectivity graph of the nodes can then be modeled as the *unit distance graph* (UDG) of the nodes, after scaling to have $r = 1$. In the UDG, two nodes are connected by an edge when their distance has length 1 or less.

- We assume that network deployment is ad hoc, so that node layout need not follow any particular geometry or topology; irregular connectivity has to be addressed. In Chapter 4 we discuss techniques for discovering this connectivity graph during network initialization.

- Nodes operate untethered and have limited power resources. Directly or indirectly, this limits and shapes all aspects of the node architecture, including the node's processing, sensing, and communication subsystems. For communication, the main consideration is that communication paths consisting of many short hops may be more energy efficient than paths using a few long hops, as discussed in Chapter 1. Also, since idling the radio costs almost as much power as transmitting or receiving, algorithms are needed that keep as many nodes in sleep mode as possible. A separate paging radio channel is sometimes advantageous, in order to avoid periodic wake-ups.

- Most lightweight sensor nodes have limited or no mobility. This makes sensor networks somewhat different from their ad hoc mobile network counterparts. If mobility is to be added, a substantially larger form-factor is needed, leading to issues akin to those addressed in distributed robotics when swarms of robots need to be controlled. Even with no mobility, sensor nodes can sleep, or fail because of power drainage or other reasons; link connectivity as well can come and go as environmental conditions vary. Thus dynamic topology changes have to be considered.

- We assume that nodes know their geographic position. Algorithms for discovering the node locations are discussed in Chapter 4.

3.2 Medium Access Control

Networking involves multiple layers in the protocol stack. Our emphasis in this chapter will be on the higher layers of the stack, those dealing with routing and application-dependent processing. Successful routing, however, also depends on the reliable transfer of information across individual physical links. Since in our setting these links are wireless and involve a shared medium, we need to say a few words about the Media Access Control (MAC) sublayer of the data link layer, at least to the extent that sensor networks may generate different MAC requirements from ad hoc wireless networks in general.

The MAC sublayer manages access to the physical network medium, and its fundamental goal is to reduce or avoid packet collisions in the medium. Several characteristics of wireless sensor networks point to the need for a specialized MAC protocol:

- Sensor networks are collaborative systems, usually serving one or a small number of applications. Thus issues of fairness at the node level are much less important than overall application performance (unlike, say, on the Internet).

- In many sensor network applications, most sensor nodes are idle much of the time. When events of interest occur and are detected, there is likely to be a flurry of activity in only some parts of the network, possibly far from where that information is needed. Because of this sporadic and episodic nature of the processing, applications must already be prepared to deal with rather large latency times. At the same time, collaboration among nodes sensing the same phenomenon can be facilitated by localized node scheduling for medium access.

- In-network processing can greatly improve bandwidth utilization.

- The assumed lack of mobility and therefore the relatively fixed neighborhood of each node can be exploited in medium access protocol design.

- As mentioned earlier, issues of energy efficiency, scalability, and robustness remain paramount. We are typically willing to compromise on many standard protocol objectives (such as fairness or latency) for the sake of prolonging network lifetime.

There are many MAC protocols that have been developed for wireless voice and data communication networks. Typical examples include the time division multiple access (TDMA), frequency division multiple access (FDMA), code division multiple access (CDMA) for wireless cellular networks, as well as carrier sense multiple access (CSMA) for the Ethernet and wireless local area networks (WLAN). So far, the most significant published work on developing a sensor network–specific MAC is the S-MAC [226] from the University of California, Los Angeles. The emerging draft standard IEEE 802.15.4 [58] is another work in support of applications low-rate personal wireless networks for monitoring and control using embedded or interfaced devices.

3.2.1 The S-MAC Protocol

The main goal of the S-MAC protocol is to reduce energy waste caused by idle listening, collisions, overhearing, and control overhead. The protocol includes four major components: periodic listen and sleep, collision avoidance, overhearing avoidance, and message passing.

Periodic listen and sleep is designed to reduce energy consumption during the long idle time when no sensing events happen, by turning off the radio periodically. To reduce latency and control overhead, S-MAC tries to coordinate and synchronize sleep schedules among neighboring nodes by periodic (to compensate for clock drift) exchanges of the nodes' schedules, so that sleep times will be synchronized whenever possible.

Collision avoidance in S-MAC is similar to the distributed coordinated function (DCF) for IEEE 802.11 ad hoc mode, using an RTS/CTS exchange. If a node loses in contention for the medium, it goes to sleep and wakes up when the receiver is free and listening again. The node knows how long to sleep, because a duration field in each packet indicates how long the remaining transmission will be. Thus overhearing avoidance is accomplished by putting nodes to sleep while their neighbors are talking to each other.

Messages are treated as logical data units passed between sensor nodes. In particular, a long message is fragmented into packets and sent in a burst with one RTS/CTS exchange to reserve this medium for the entire message. This saves repeated RTS/CTS overhead and reduces overall message-level latency. It does mean that a short message may have to wait a long time while a long message finishes transmission, but as we remarked, node-level fairness is not so important in sensor networks.

Because of such measures targeting improved energy efficiency, the energy consumption gap between 802.11-like protocols and S-MAC becomes significantly wider as the message interarrival period increases. Therefore, for an ad hoc sensor network with nodes remaining largely inactive for long times, S-MAC has obvious advantages over the 802.11 MAC in supporting typical sensor network applications today.

3.2.2 IEEE 802.15.4 Standard and ZigBee

The IEEE 802.15.4 standard defines both the physical and MAC-layer protocols for most remote monitoring and control, as well as sensor network applications. ZigBee is an industry consortium with the goal of promoting the IEEE 802.15.4 standard. ZigBee ensures interoperability by defining higher network layers and application interfaces. The low-cost, low-power features of 802.15.4 are intended to enable the broad-based deployment of wireless networks able to run for years on standard batteries, for a typical monitoring application. It is optimized for low data throughput (up to 115.2 Kb/s), with simple or no QoS support. Both star or peer-to-peer network

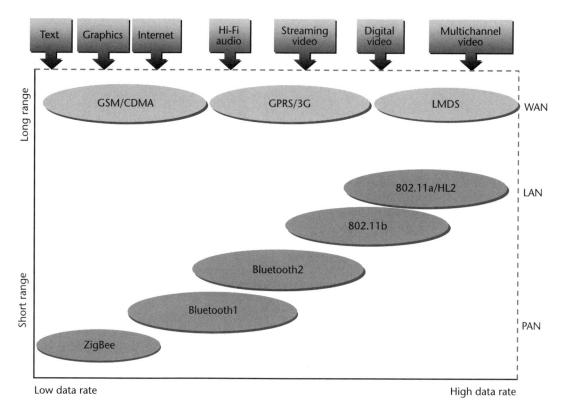

Figure 3.1   A bird's-eye view of wireless technologies, according to data rate and range.

topologies are supported. Unlike S-MAC, 802.15.4 achieves its power efficiency from both the physical and MAC layers. The duty cycle of communications in an 802.15.4 network is expected to be only around 1 percent, resulting in very low average power consumption. However, it is also up to the higher protocol layers to observe the low duty cycle requirement.

Figure 3.1 shows the landscape of different wireless technologies.

3.3 General Issues

Two essential characteristics of current sensor networks, as mentioned repeatedly, are that the nodes have only modest processing

power and memory, and that network links and nodes can come and
go. When it comes to packet routing, this makes classical routing protocols, such as those used to implement TCP/IP addressing, much less
attractive in the sensor network context. Such protocols require large
routing tables (one entry for every possible destination) to be maintained at each router. In the sensor net setting, every node has to
act as a router if need be. Furthermore, proactive protocols of the
distance vector (DV)-type need to broadcast to the entire network
topology changes when these are locally detected. Since topology
changes are a relatively frequent occurrence in the sensor network
case, this would use a substantial fraction of the available bandwidth
and energy to accommodate a flood of update messages. Link state
(LS)–type protocols only broadcast topology changes to neighbors,
but because of that they converge slowly; furthermore, in the sensor
net setting, synchronization difficulties can arise if multiple topology
changes are propagating simultaneously through the network.

Several strategies have been suggested to mitigate the demands of
these dynamic changes on the network:

- The frequency of topology updates to distant parts of the network
  can be reduced, as in *fisheye state routing* [178].

- Reactive protocols can be used instead, constructing paths on
  demand only. Examples include *dynamic source routing* (DSR) [108]
  as well as *ad hoc on demand distance vector routing* (AODV) [180].

However, it is even more attractive to focus on local stateless algorithms that do not require a node to know much more beyond its
immediate neighbors. This is the approach we will develop in the
chapter.

But perhaps the most profound difference between sensor networks and other types of network, whether wired or wireless, comes
from the fact that the classical separation of address and content in
a packet is no longer viable in the sensor net setting [66, 95]. From
the point of view of the network as a system, what really matters
is the information held or obtainable by nodes, and not the nodes

themselves. After all, the nodes are fragile objects that can easily be destroyed; and whatever information can be sensed from one node usually can also be sensed from another nearby node. This gives rise to a data-centric view of the network [66], where routing decisions are to be made based not on destination addresses, but rather on destination attributes and their relation to attributes of the packet content. Thus information providers and information seekers need to be matched using data attributes and not hard network addresses. Examples of such attributes may include

- a node's location

- a node's type of sensors

- a certain range of values in a certain type of sensed data

New mechanisms are necessary to help route packets using destination attributes and perform the necessary matchmaking to bring together those nodes having information with those desiring to use it. Note that both the pull and push models apply here. The network must be both a database that can be queried about the world state (pull), but also an entity that can actively initiate an action when something of interest is sensed (push). How this matchmaking can happen is addressed in Subsections 3.5.1, 3.5.2, and 3.5.3.

3.4 Geographic, Energy-Aware Routing

In this section we focus on routing protocols whose aim is to deliver packets to nodes or areas of the network specified by their geographic location. As mentioned earlier, for sensor networks, the most appropriate protocols are those that discover routes on demand using local, lightweight, scalable techniques, while avoiding the overhead of storing routing tables or other information that is expensive to update such as link costs or topology changes. This lack of global information creates challenges in discovering paths, especially paths that are both time- and energy-efficient for the particular transmission desired, as well as energy-aware in terms of load-balancing utilization across the entire network.

The assumptions we make are that

- all nodes know their geographic location (see Chapter 4);

- each node knows its immediate one-hop neighbors (those within its radio range);

- the routing destination is specified either as a node with a given location, or as a geographic region (more details later); and

- each packet can hold a bounded ($O(1)$) amount of additional routing information, to help record where it has been in the network.

For a general survey of position-based routing in mobile ad hoc networks the reader is referred to [223].

Let V be the set of nodes and with $|V| = n$. Let G be the connectivity graph of V, where there is an edge from node x to node y if and only if x can talk directly to y. Since we know the geographic locations of the nodes, it is convenient to think of V as a set of points in the plane. In the simplest possible model, we ignore all radio propagation effects such as obstacle or multipath interference, anisotropy or asymmetry of transmission, and the like and make all nodes have the same transmission range which we set to be equal to 1, by rescaling if necessary. The graph G then is just the undirected *unit-distance graph* (UDG) of the points V mentioned earlier: there is an edge between any two points whose distance is less than or equal to 1. Depending on the placement of the nodes, G can vary from the complete graph on n vertices (when all the nodes are very close), to n isolated points (when the nodes are far apart). In most cases of interest, each node can talk to a small number of immediate neighbors and the entire graph G is connected.

We should remark that the equal-transmission-range assumption we make is for simplicity of exposition only. Even if the physical maximum transmission range of each node is the same, it may be more economical to use a smaller transmission radius for nodes in areas of high node density—without sacrificing adequate network connectivity. This is the issue of *topology control*, discussed in Chapter 4; for more information see page 128.

3.4.1 Unicast Geographic Routing

We start with the simplest case, where we wish to route a message to a single node whose coordinates are given. Let s be the message source node and t the message destination node. Denote by x the node currently holding the message on its way from s to d. The main task of the routing protocol is to choose a neighbor y of x to pass on the message. If we assume that no precomputation has been done to aid routing based on the topology of the network, a couple of obvious locally optimal strategies suggest themselves:

greedy distance routing: Among the neighbors y of x closer to d than x, pick the one closest to d [22].

compass routing: Among the neighbors y of x that make an angle $\angle dxy < \pi$, we pick one that minimizes the angle $\angle dxy$ [122].

See Figure 3.2. In both cases we resolve ties arbitrarily.

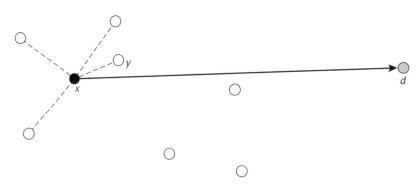

Figure 3.2 Node x selects node y as the locally optimal neighbor to hand off a packet whose destination is d.

While clearly simple and easy to implement, these two protocols raise several concerns. The intuition behind both is to have the message packet make progress toward its eventual destination. However, this may not always be possible. For example, what if, in the greedy distance routing protocol, all neighbors of x are farther from d than x itself? This can easily happen, as Figure 3.3 shows; such situations commonly arise when the packet on its way from s to d encounters a "void" or "hole" in the network. In this case, the message packet will be stuck at x. (If we allow the packet to backtrack from x, we can easily create stuck situations where this greedy rule forces the packet to oscillate between only two nodes x and y.) Analogous situations can happen with compass routing.

The only way to guarantee that these situations do not arise is to make some special assumptions about the connectivity graph G that avoid connectivity "holes" at whose boundaries packets can get stuck at local minima. For instance, suppose that every node encountered by the packet has a set of neighbors that cannot be confined in a single halfspace, a situation we would certainly expect to be true if we have a dense configuration of sensor nodes (except near the

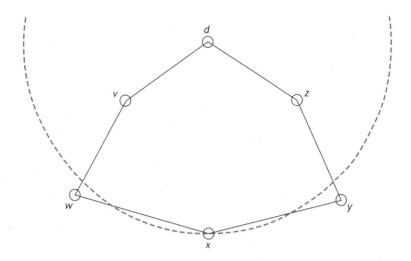

Figure 3.3 A greedy forwarding strategy can get stuck at node x: both of x's neighbors are farther from the destination D than x itself.

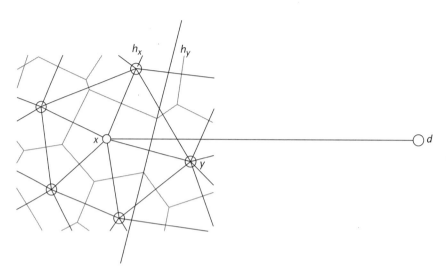

Figure 3.4   In the Delaunay triangulation, node *x* must have a neighbor *y* closer to *d* than itself.

boundaries of the sensor field). Even then, though the packet cannot get stuck at a single node when following the compass protocol, it may still get trapped going around a cycle [122] and never reach the destination *d*. Only global assumptions about *G* seem to be sufficient to guarantee delivery. For instance, if *G* contains the *Delaunay triangulation* [51] of *V*,[1] then the greedy distance protocol always works. To see this, consider how the segment *xt* traverses the *Voronoi diagram* [51] of *V*. When the segment leaves the cell of node *x*, it enters the cell of another node *y*—see Figure 3.4. It is easy to see that *y* is both closer to *d* than *x* and a Delaunay neighbor of *x* [22]. Since only a finite number of points are involved, the process must terminate with the packet arriving at *d*. Compass routing has guaranteed delivery when *G* is the Delaunay triangulation of *V* as well [122].

---

1 The Delaunay triangulation and its dual, the Voronoi diagram, are two basic structures in computational geometry, discussed at length in almost any computational geometry textbook; see, for example [51]. Briefly, the Voronoi diagram of *n* sites in the plane is defined as the partition of the plane into regions, according to the nearest site; the Delaunay triangulation of the same sites is the triangulation, dual to the Voronoi diagram, satisfying the condition that the circumcircle of every triangle is free of all other sites.

In practice, however, unless the spacing of the nodes is very dense, it is unlikely that all Delaunay edges will be in the unit-distance graph of V.

Guaranteeing delivery when a path exists is not the only desideratum for such routing protocols. Though it is easy to tell when we are stuck at a local minimum, knowing that we are trapped in a cycle may be more challenging, unless we are allowed to leave some markers behind in the network or carry some information along in the packet about where it has been. Thus the task of discovering that no path from s to d exists may be nontrivial as well.

In many situations, we also want the path we find to be optimal, or at least close to optimal, among all possible paths between s and d in G. There are some difficulties in agreeing on what *optimal* means, as different criteria may be at odds with one another. For instance, minimum delay may favor paths with the fewest hops, while minimum energy may favor paths with many short hops [88]. In general, we can express the cost $c(\pi)$ of a path π as

$$c(\pi) = \sum_{e \in \pi} \ell^k(e).$$

Here $\ell(e)$ denotes the length of edge e in G, and the usual values for k are between 0 and 5. The value $k = 0$ captures the hop length of the path (number of edges) and is a measure of delay; the value $k = 1$ is the Euclidean path length; and, finally, $k \geq 2$ captures the energy of the path, depending on the attenuation model used.

Fortunately, these multicriteria difficulties do not arise if we assume that, before routing is done, a clustering algorithm is applied to the node locations to aggregate nodes into clusters whose diameter is comparable to the node communication radius as suggested in Chapter 4. Each cluster elects a cluster-head node, and routing is done only among the cluster-heads (the remaining nodes always route packets through their cluster-heads) [77]. This is advantageous for a variety of reasons, including the possibility of using simpler communication protocols within a cluster, recycling of resources (such as frequency assignments) among disjoint clusters, and saving power.

A particular consequence of clustering is that we can now assume that there is a minimum separation among the cluster-heads comparable with the radio communication distance and that as a consequence each cluster-head talks directly to only a bounded ($O(1)$) number of other cluster-heads. In that situation, the path measures given above for different exponents d all become equivalent to within a constant factor, and thus the differences among them can be ignored for our purposes.

Planarization of the Routing Graph

Surprising as it may seem, a way to get protocols that guarantee geographic packet delivery is to *remove* some edges from the connectivity graph G so as to keep the same connectivity but make the graph planar. The advantage is that on planar graphs, there are simple exploration protocols based on the ancient idea of traversing a maze by keeping one's right hand against the wall.

Suppose that we draw G as a graph in the plane by connecting via a straight line segment every two nodes defining an edge of G. Suppose further that no two of these segments intersect, except at endpoints. Then G is planar and what we have drawn is a planar subdivision corresponding to the standard embedding of G in the plane. Each node is assumed to know the circular order of its neighbors in this embedding. Suppose also that all faces of G, except the outer infinite face, are convex; the latter is assumed to be the complement of a convex polygon. Then the line segment connecting the source node s to the target node d crosses a series of convex faces. In this case, we can route from s to d as follows:

- *Convex perimeter routing*: Start in the face of G just beyond s along sd and walk around that face (say, counterclockwise). Stop if d is reached or if the segment sd is about to be crossed. In the latter case, cross over into the next face of G along sd and repeat the process.

See Figure 3.5 for an example. A simple modification of this idea yields a protocol with guaranteed delivery for arbitrary planar

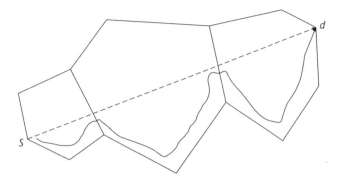

Figure 3.5 The packet gets from node *s* to node *d* by walking around the faces of the planar subdivision.

graphs [122]—a protocol we will henceforth refer to simply as *perimeter routing*. The main difference from the convex case is that the segment *sd* may intersect a face more than twice and the algorithm needs to determine all face boundary crossings with *sd* and select the one farthest from *s*. All this can be done while using only bounded storage in the packet. Note that if the destination *d* is not reachable from *s*, then the perimeter protocol will loop around an interior or exterior face loop of the planar graph *G*. Thus, to defect this, the packet must remember the first edge traversed when it starts going around a new face.

A variant of perimeter routing is the *other face routing* (OFR) protocol of [127]. In OFR, after we go around a face *F*, we continue into a new face *F'* not from the farthest with *F*, but instead from the vertex of *F* closest to *d*.

In general, not much can be said about the quality of paths discovered by these protocols. We can easily construct an example where, had we chosen to go around the faces clockwise, we would have found a path with $O(1)$ hops, while by going counterclockwise, our paths ends up having $\Omega(n)$ hops—which is no better than flooding the network.

A way to avoid such bad paths is to limit how far away from the segment *sd* the path is allowed to veer. Note that if the length of the optimum path is *L*, then every vertex on the optimum path must lie inside an ellipse with foci s and d defining the locus of points whose sum of distances to s and d is L. If we knew L, then we could run perimeter routing, but not allow ourselves to go outside this ellipse. If we hit the ellipse boundary while going around a face, we would just turn around and walk clockwise around the face instead. It is not hard to see that the process would find a path from s to d, but in the process it could visit many or all the nodes of G inside the ellipse. If we made use of the cluster-head idea discussed earlier, then we would know that cluster-heads have bounded density and therefore the number of nodes inside the ellipse is $O(L^2)$. Of course we do not know L, but we can easily get around this by guessing L. We can start with a small guess and double it every time we fail to reach d with our current guess, until d is reached. This defines adaptive versions of the above protocols, called *adaptive perimeter routing* and *other adaptive face routing* (OAFR) [127]. The total work and number of nodes on the eventual path will still be $O(L^2)$, which is shown to be asymptotically optimal in [126].

Although it is possible, it is unlikely that the connectivity graph G will just happen to be planar. Therefore, we need to discuss techniques for removing edges from G to make it planar. In doing so, there are two considerations to keep in mind:

- The planarization algorithm should be efficiently executable in a distributed fashion.

- The quality of paths in the sparsified planar graph should not be too much worse than in the original connectivity graph.

A way to formally capture the last property is via the notion of graph spanners [179] on edge-weighted graphs. We say that a subgraph H of G is a *spanner* for G if H is a graph on the same vertex set as G, is missing some of the edges of G, yet there is a constant α, $\alpha \geq 1$, such that for any pair of vertices u and v the length of the shortest path between u and v in H is at most α times that of the length of the shortest path between u and v in G. The constant α is called the *stretch factor* of the spanner H with respect to G.

A number of the standard geometric graph constructions for a set of points V can be used to define planarizations of the graph G. It is appealing to use local constructions, in which an edge xy is introduced for nodes $x, y \in V$, if a geometric region (the *witness* region) around the edge xy is free of other nodes. The most common such constructions are:

- *the relative neighborhood graph (RNG)*, where the edge xy is introduced if the *lune* [intersection of the circles centered at x and y with radius the distance $d(x, y)$] is free of other nodes

- *the Gabriel graph*, where the edge xy is introduced if the circle of diameter xy is free of other nodes.

Both of these graphs are known to be planar and, of course, the Gabriel graph is a supergraph of the RNG [107]; see Figure 3.6 for an example. In our case, only edges of length less than or equal to 1

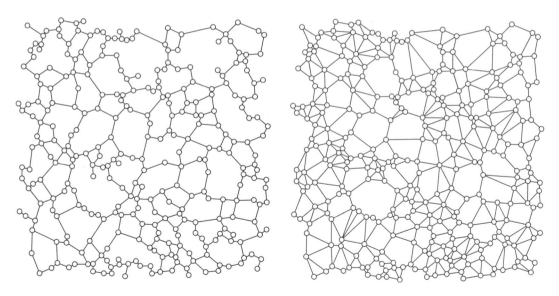

Figure 3.6   An example of the relative neighborhood graph (RNG) on the left, and of the Gabriel graph on the right.

would be considered for inclusion in H. We will denote the resulting subgraphs by RNG(G) and Gabriel(G). The good news is that both can be computed by simple local algorithms (assuming that a pair of nodes within distance 1 can talk to each other); the bad news is that it is known that neither is always a spanner for G [23]—they are just too sparse for that.

The Delaunay triangulation is known to be both planar (or course, being a triangulation) and a spanner [54, 113] (best-known stretch factor = 2.42) of the complete Euclidean graph on V. However, the Delaunay triangulation is globally defined. For a local version, we can consider a subgraph U consisting only of Delaunay edges of V whose length is at most 1. This subgraph can be shown to be a spanner of G [77, 133]. It is difficult to construct exactly this subgraph U of the connectivity graph G through localized algorithms, but recently it was shown how to build certain subgraphs of G containing U that are still planar. Thus both the *restricted Delaunay graph* (RDG) [77] and the *localized Delaunay triangulation* (LDel) [133] satisfy the two criteria above for desirable planarizations of G.

The above algorithms all depend on sufficiently accurate knowledge of the positions of the nodes. As discussed in Chapter 4, significant errors are possible in the node localization process. Even if only one node has an erroneous location, the above planarization algorithms can fail. An incorrect edge removal may disconnect the network, while failure to remove an edge that should have been removed may lead to crossing edges in face cycles that cause face routing to fail. These and more types of failures are discussed in [207], where a number of simple fixes are also proposed, especially for the disconnection problem. In the RNG and GG planarizations, these boil down to not removing an edge of U unless both of its endpoint nodes can see (communicate with) the third node whose presence in a witness region is the cause for the removal of that edge.

On the other hand, even if exact positions of most of the nodes are not known, it may be possible to assign to nodes *virtual coordinates*, so that geographic routing can still work. Such virtual coordinates can be based on the known positions of some nodes and the overall connectivity graph of the network, as discussed in [189].

Greedy Perimeter Stateless Routing

While greedy distance protocols are extremely simple, they can get stuck in local mimima. In contrast, protocols based on planar subgraphs can provide guaranteed delivery, but require both preprocessing and a significantly more complex routing algorithm. In many situations where sensor nets are deployed, we can expect that node distribution will be fairly uniform over large areas of the sensor field, but there will be a number of holes or uncovered areas, due to obstacles, node depletion, and the like. Thus it makes sense to consider merging these two approaches, to get the best of both worlds.

The *greedy perimeter stateless routing* (GPSR) protocol [112] is such a mixture. A planar subgraph H of G is computed during a preprocessing phase. Normally GPSR defaults to greedy distance routing, as discussed earlier. When greedy routing gets stuck, then the perimeter protocol is invoked, but only until a node is encountered that is closer to the destination than the node where the greedy distance protocol was stuck. Then GPSR continues with the greedy distance protocol again from there. The original paper [112] suggested RNG(G) and Gabriel(G) as the planarizations of G. Later, another paper [77] argued that the RDG produces better paths in both theory and practice.

Incidentally, just because the routing subgraph H to be used may be a spanner and therefore contains paths nearly as good as those in G, there is no guarantee that a stateless protocol will find those. Actually, even if the perimeter routing part of GPSR is never invoked, there is no guarantee that the path produced by GPSR will be close in length to the optimal path from s to d. This is an area in need of further research, as currently such path-length guarantees can be given only if the graph G satisfies some global conditions (such as being the full Delaunay triangulation of V), and the required protocol (though still conforming to assumptions) needs to be considerably more complex [23].

3.4.2 Routing on a Curve

Another form of geographic routing, applicable in a number of dense node coverage settings, is to specify an ideal curve that a packet should follow, as opposed to the packet's final destination [167]; this has been called *trajectory based forwarding* (TBF) [170]. The curve is specified analytically and its description is carried along on the packet, as the latter hops from node to node following a trajectory that approximates the ideal curve. This can be useful in situations where such curves correspond to natural features of the environment in which the nodes are embedded. For example, in a sensor network deployed over a geographic terrain, we may want to do selective broadcasting to nodes that follow a specific road or topographic contour. Routing on curves can simulate flooding from a source, by propagating messages along radial spokes emanating from the source. It can also be used to implement rumor routing (discussed in Section 3.5.2), a method for bringing together information seekers and information providers in a sensor network.

Like routing to a destination, routing on a curve can be implemented in a local greedy fashion, requiring no preprocessing or global knowledge of the topology of the network. It is convenient to assume that the curve is specified in parametric form $C(t) = (x(t), y(t))$, as then the parameter t provides a convenient mechanism for naming the points of the curve. Although we can think of t as time, it is important to keep in mind that t need not be related to real time as the packet advances along the curve. A packet arriving at a node v during curve routing specifies the analytic form of $C(t)$, as well as the time (parameter) t_v of the closest approach of the curve to v. The node v then decides to which of its neighbors it will forward the packet next, making use of its knowledge of both the curve trajectory and the neighbor positions. In this process, the node v estimates the distances from its neighbor nodes to the curve (the so-called residuals), as well as the times of closest approach to each of them within its communication region. Again, different policies are possible [167, 170]. We may choose the neighbor w of smallest residual whose closest approach time t_w is larger than t_v, if the goal is to stay as close to $C(t)$ as possible. Or we may choose the neighbor w whose residual is within a certain deviation tolerance and whose time of closest approach t_w is the largest, if the goal is to advance as quickly as possible while staying near the curve.

Note that similar ideas can be used to send a packet along a geometrically defined tree, thus achieving certain kinds of multicasting. It is also worth observing that such geometric route specifications completely decouple the route description from the actual locations of the network nodes involved in transport. This makes such schemes very robust to network or link failures resulting in topology changes, as long as an appropriate level of node density is maintained along the packet trajectory.

3.4.3 Energy-Minimizing Broadcast

We have repeated several times the need to be energy-aware in communications involving a sensor network. Two aspects of the energy cost in a sensor network make it challenging to reason about optimizing energy:

- Multihop communication can be more efficient than direct transmission.

- When a node transmits, all other nodes within range can hear.

These are both a consequence of the fact that nodes communicate using radio links, where the signal amplitude drops with distance according to a power law of the form $O(1/r^\alpha)$, where typically $2 \le \alpha \le 5$.

To illustrate these issues, we discuss in this section the *minimum-energy broadcast* problem. In this problem we have a network of n nodes with known positions that communicate with each other via radio transmissions. We assume a signal attenuation law of the form $O(1/r^\alpha)$ as previously, and that nodes have enough power that, if need be, even the most distant pair of nodes can communicate directly. We have a special source node s that wishes to broadcast a message to all the other nodes in the network. Multihop communication is allowed, and our goal is to find a schedule of broadcasts and retransmissions that allows the message to reach all nodes while minimizing the total energy expended. The analysis that follows is overly simplified for pedagogical reasons. In particular, we ignore the energy spent in transmitter and receiver electronics, focusing only on the energy used to communicate across the medium. Delay and interference are also unaccounted for in this formulation. For a discussion of these issues, see [161] and the references therein.

If we had point-to-point communication among nodes (with the same energy cost as in our problem), then what we are asking is to compute the *minimum spanning tree* (MST) of the full communication graph on the n nodes. This is a classical optimization problem for which numerous fast polynomial algorithms exist [46] (nearly linear in both the graph and geometric settings—in the former case, linear in terms of the graph size means quadratic in terms of the number of nodes). The broadcast nature of radio transmissions, however, changes the problem, as node u transmitting to reach another node v also reaches all nodes closer than v at no extra cost. This is called the "wireless multicast advantage" in [225].

In the minimum-energy broadcast problem, we ask for the construction of a min-cost broadcast tree. In this tree the leaves are passive receivers, while the internal nodes are relays, retransmitting the message received from their parents. Each node retransmits with just sufficient power to reach its farthest child (and therefore all of them), when it receives the message from its parent. Consider the simple example of three nodes shown in Figure 3.7. The source s wishes to broadcast a packet to nodes v_1 and v_2, having distances r_1

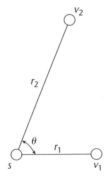

Figure 3.7 A simple broadcast scenario: Node s wishes to reach nodes v_1 and v_2.

and r_2 to s, respectively. Let θ be the value of the angle $\angle v_1 s v_2$. In this example there are two broadcast strategies:

- s transmits to v_2; both v_1 and v_2 are reached, or

- s transmits to v_1 and then v_1 transmits to v_2.

The energy needed for the first strategy is r_2^{α}, while that for the second is $r_1^{\alpha} + r_{12}^{\alpha}$, where r_{12} denotes the distance between v_1 and v_2. When $\alpha = 2$, it is easy to see that the first strategy is advantageous if and only if $r_1 > r_2 \cos \theta$. A little more algebra shows that this condition becomes

$$(1 + x^2 - 2x \cos \theta)^{\alpha/2} > x^{\alpha} - 1,$$

for general α, where $x = r_2/r_1$. A numerical investigation shows that if $\alpha \geq 3$ (which is true in many realistic situations), then using the first strategy when $\theta \geq 90 \deg$ and the second otherwise gives very nearly optimal results. In general, the higher the α, the more we want to use the shortest available links.

When n is large, the number of different strategies to consider becomes exponential, and indeed it has been shown that the minimum-energy broadcast problem for both general and geometric graphs is NP-complete [29]. This is a very different situation from the case of wired, point-to-point links. Because of this, a number of authors have investigated approximation algorithms that try to find a broadcast tree that is within a constant factor of the optimum energy [225]. A possible approach falls back on the use of the MST, which can be computed fast. The MST of the n nodes, when rooted at the source s, becomes a broadcast tree for solving the minimum-energy broadcast problem. How well does this do? It can be shown that in the worst case, this MST-based algorithm gives a broadcast tree whose total energy usage is between 6 and 12 times the minimum needed [221]. Thus it is a constant-factor approximation algorithm, though not a very good one.

The key idea in [221] that establishes the constant-factor approximation is a proof that the minimum energy required for the broadcast problem can be bounded from below by a constant times the energy cost of the MST-based broadcast tree. This argument is based on two key observations.

- A traditional MST T minimizes $\sum_{e \in T} |e|$ (the sum of the edge lengths) over all possible spanning trees T of the n nodes. However, it is well known that the combinatorial structure of the MST depends only on the ordering of edge weights, and not the actual weights. Raising lengths to a positive power preserves all comparisons, since all weights in consideration are nonnegative. Thus the traditional MST also minimizes $\sum_e |e|^\alpha$ over all spanning trees for all α of interest.

- It is shown in [221] that $\sum_{e \in \mathrm{MST}} |e|^\alpha = O(d^\alpha)$ for any $\alpha \geq 2$, where d denotes the radius of the set of nodes (smallest r such that a ball of radius r around some node covers all the rest).

Let B denote an optimal broadcast tree. For each relay node $p \in B$, let T_p denote a minimum spanning tree of itself and its children in B. We denote the radius of the nodes in T_p by r_p. Then we have

$$\sum_{e \in T_p} |e|^\alpha = O(r_p^\alpha),$$

or equivalently

$$r_p^\alpha = \Omega \left(\sum_{e \in T_p} |e|^\alpha \right).$$

Summing over all relay nodes p, we get that the total energy used by B is proportional to

$$\sum_p r_p^\alpha = \Omega \left(\sum_{e \in U} |e|^\alpha \right),$$

where U denotes the union of all the edges in the local MSTs of the relay nodes p. If we use a global MST T instead of U in this sum, we only make the total smaller by the first observation. Thus the claim that the cost of B is lower bounded by a constant factor times the broadcast cost of T follows.

A more practical method—without guarantees, however—is the *broadcast incremental power* (BIP) algorithm of [224]. That method works like Prim's MST algorithm [46], adding nodes to the broadcast tree one at a time, starting with the source. At each stage, a node is added that has the minimal incremental cost to be connected to one of the already selected nodes. For example, in Figure 3.7, node v_2 may be added to the tree by having node v_1 broadcast to it, or by having node s increase its power to reach v_2 as well, according to whichever alternative is cheaper. The total power used is the sum of the maximum broadcast powers used for s and all other relay nodes.

3.4.4 Energy-Aware Routing to a Region

Instead of broadcasting to all nodes in the network, a more common situation is the wish to reach all nodes in a certain geographic region. For example, we may be concerned only about seismic vibrations near a downtown area with high buildings. The problem of routing a message to a region combines two of the problems we have discussed so far: unicast geographic routing (Section 3.4.1) and energy-minimizing broadcast (Section 3.4.3): first we have to get the message to the region of interest, and then we have to distribute it to all the nodes in the region. Energy considerations matter in the first phase as well. Protocols such as GPSR, which circumvent holes in the network by routing along nodes on hole boundaries, may quickly deplete these nodes if lots of traffic is routed around the hole. Depleting paths is, of course, always undesirable, but its impact can be mitigated if there are other fairly good paths nearby. Depleting hole boundaries, however, can have more disastrous effects, because depleting effectively makes the holes grow larger, and eventually several small holes can merge into a giant hole that causes significant and unnecessary delays in message delivery.

In this section, we discuss a protocol named GEAR which stands for *geographical and energy-aware routing*, as presented in [229], whose goal is to efficiently route a message to a geographic region while at the same time performing some load-balancing on the nodes used and thus avoiding energy depletion. The GEAR protocol operates in two phases:

- deliver the packet to a node in the desired region

- distribute the packet within the region

For simplicity, we assume that the geographic destination region R is specified as a rectangle aligned with the axes (for more complex shapes, this approach can be used with a bounding box of the shape; the true shape can be encoded in the packet and then used to filter out unwanted nodes during the distribution phase).

During the delivery phase, GEAR behaves like a unicast protocol routing the message to the centroid d of the region R, except that it considers the energy resources of each node as well. In addition, GEAR does not use any specialized subgraphs to route around holes. Instead, it relies on a generic A^*-type search algorithm, the *Learning Real-Time A^** (LRTA*) algorithm of Korf [120]. As the search visits various nodes, each visited node x maintains its learned cost $h(x, d)$ to the destination d; this cost balances the length of the best path from x to d while favoring nodes whose energy deposits are still plentiful. When a node y is first encountered during the search, its learned cost is set to

$$h(y, d) = \alpha \ell(\overline{yd}) + (1 - \alpha)E_y,$$

where $\ell(\overline{yd})$ is a normalized Euclidean distance from y to d and E_y is a normalized form of the energy already consumed from node y. The mixing factor α indicates the relative importance of shorter path length versus less energy depletion.

The GEAR protocol works by performing the following steps at each visited node x:

- If x has neighbors closer to d than x in both the Euclidean sense and the learned cost sense, choose among these neighbors the one with the smallest learned cost and pass the packet to that neighbor.

- Otherwise, forward the packet to the neighbor of minimum learned cost.

Furthermore, each node x periodically broadcasts its learned cost $h(x, d)$ to its neighbors. When x itself receives such an update from a neighbor y, it performs a classical edge relaxation step, updating its own learned cost to

$$h(x, d) \leftarrow \min \{c(x, y) + h(y, d), h(x, d)\},$$

where $c(x, y) = \alpha \ell(\overline{xy}) + (1 - \alpha/2)E_x + (1 - \alpha/2)E_y$. When the relaxation step lowers the learned cost $h(x, d)$, then effectively the node x has discovered a better path to d going through neighbor y. However, if the current value of $h(x, d)$ was obtained through an earlier communication from y, then $h(x, d)$ is always updated to $c(x, y) + h(y, d)$, so that invalid paths do not persist.[2]

For an example of how this works, consider the small network in Figure 3.8. Assume all shown vertical and horizontal edges have length 1 and all diagonal edges length $\sqrt{2}$. Assume also that all live nodes have the same energy left. If node s wishes to send a packet to node d, in the absence of any other information, it will choose to send it next to neighbor b, the one geometrically closest to d. Node b, however, happens to sit at a hole boundary formed by depleted nodes and has no live neighbor closer to d than itself. In this case, GEAR will choose to send the packet to node c, whose geometric distance

---

2 Note that learned costs can both increase (because of node depletion) and decrease (because of better paths being discovered).

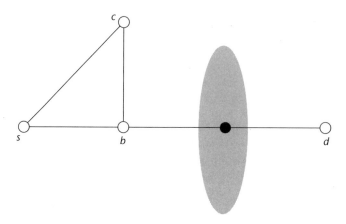

Figure 3.8   A routing example for the GEAR protocol. Node *s* wishes to route a packet to node *d*, but there is a network hole in between.

to *d* is $\sqrt{5}$. In the process, *b* will also update its learned cost to *d* to be $1 + \sqrt{5}$, which is larger than that of *c*. Thus when *s* wishes to send another packet to *d*, it will choose neighbor *c* over neighbor *b*. The completeness of LRTA* implies that when a valid path from *s* to *d* exists, it will always be found eventually in this fashion. However, no performance guarantees on the quality of the actual path used or the convergence rate can be given.

Even though there are no theoretical optimality guarantees and even the convergence of the path to a local minimum can be slow (quadratic in the network size), GEAR still behaves well in practice and nicely blends the heuristic of choosing the neighbor closest to the packet destination with that of choosing the neighbor with the most remaining energy. As compared to GPSR, in experiments described in [229], GEAR provided a 50 to 100 percent increase in the number of packets transmitted before network partition, and a 50 percent increase in the number of connected pairs of nodes after network partition. The average increase in the path length taken by a packet due to load balancing was between 25 and 45 percent.

The second phase of GEAR deals with packet distribution within the destination region *R*, once the packet has reached a node *x* in the region. One of two different strategies can be used here. *Recursive geographic forwarding* refers to a process of recursively partitioning the region R into quadrants and then forwarding the packet from x to the centroid of each quadrant. *Restricted flooding* refers to initiating a broadcast flooding process from x that is clipped at the boundary of R. While restricted flooding has the "wireless multicast advantage" mentioned earlier, it may still prove wasteful when the region R is densely populated by nodes and the same packet reaches a node multiple times. The issue of determining the best strategy for distribution within R needs further study.

3.5 Attribute-Based Routing

The previous section focused on routing algorithms based on prior knowledge about the geographic destination of a packet. In more general settings of data-centric routing, however, we cannot assume that we know either the network address or the geographic location of the node we wish to communicate with. Geographic location services can be used to map from node IDs to locations, as discussed in Section 4.4.4. In this section, we present more general protocols, whose goal is to enable nodes desiring certain types of information and nodes having that information to find each other in the network and maintain good communication paths between them. To perform this kind of matching, we will assume that data is described by attribute-value pairs that characterize the information that a node holds or seeks. For instance, a node tasked to look for animals in an environmental monitoring setting might detect a horse, based on its sensor readings (the details of how this is done do not concern us here). At that point, this node might generate an attribute-value event record of the following type:

```
type = animal          // named record type
instance = horse       // instance of this type
location = [89, 154]   // location of horse
time = 02:45:23        // time of detection
```

Each line in this record is an (attribute, value) pair. Correspondingly, a node seeking information about horses in a certain region might create an information request record of the form:

```
type = animal           // named record type
instance = horse        // instance of this type
rect = [0, 200, 0 200]  // a spatial range of interest
```

This record specifies a rectangular range indicating that the request-ing node is interested in horse detections in the specified rectangle (here, for simplicity, taken to be parallel to the axes and specified by an x and y interval). In general, ranges can be specified for any type of continuous value that can be stored in a record, thus allowing general range queries as information requests.

We now discuss a number of techniques aimed at allowing the network to discover which event records and information requests match and maintain communication paths between information sources and sinks.

3.5.1 Directed Diffusion

Directed diffusion [104] is a very general approach toward problems of this type. Nodes requesting information are called *sinks*, while those generating information are called *sources*. Records indicating a desire for certain types of information are called *interests*. Interests are propagated across the network, looking for nodes with match-ing event records. Key to directed diffusion is the assumption that interests are persistent—that is, if a source has information relevant to a sink, then the sink will be interested in repeated measurements from that source for some period of time. A typical interest record contains an interval attribute field, indicating the frequency with which the sink wishes to receive information about objects matching the other record attributes. This longevity of communication pat-terns allows the directed diffusion protocols to learn which are good paths between sources and sinks and amortize the cost of finding these paths over the period of use of the paths (the period of validity of an interest is encoded in its duration attribute).

Sinks generate information request tasks, or interests, that *diffuse* through the sensor network. All nodes track the unexpired interests they have seen, but not their originating sinks. Each node maintains an interest cache, with an entry for each distinct interest the node has seen and whose duration has not yet expired; the node knows only from which neighbor(s) this interest came. Each node in turn may choose to forward an interest to some or all of its neighbors, and so on. An essential component of directed diffusion is the use of *gradients* associated with each interest cache entry, used to direct and control information flow back to the sink, as we will see. Since the network is not assumed to be perfectly robust, periodically each sink rebroadcasts its interest message. A monotonically increasing timestamp attribute in each interest record is used to differentiate these repeated broadcasts from earlier versions.

A gradient is typically derived from the frequency with which a sink requests repeated data about an interest, as mentioned earlier, and indicates the frequency of updates desired and the neighbor (direction) to which this information should be sent. An elegant aspect of directed diffusion is the way in which gradients can be manipulated to reinforce good information delivery paths and disable unproductive ones. Note that it is perfectly possible for interest diffusion to create two neighboring nodes having gradients toward each other for the same interest. This multiplicity of gradients does not create persistent data delivery loops, as we will see, yet allows for the quick reestablishment of information delivery paths when nodes or links fail.

To understand how directed diffusion works, let us consider a simple example. Let us say that a sink is interested in horse detections in a certain small area containing a single sensor node. Since a route to that area is not known to the sink, the initial interest will have to flood the network. For this reason, the initial requested data rate is set to an artificially low value, so as to avoid excess traffic along multiple return paths. The interest diffusion now proceeds away from the sink, until a source with horse detections in the area of interest is reached

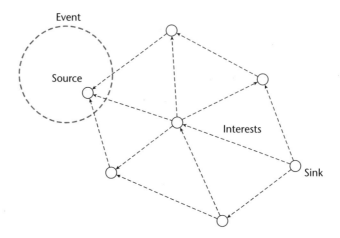

Figure 3.9   The directed diffusion algorithm propagates interests from a sink, until an appropriate information source is reached (adapted from [104]).

(see Figure 3.9). When the matching event record is found, the source computes the highest outgoing event rate among all its gradients for that interest. The source node then tasks its sensor subsystem to generate event samples at this maximum data rate and sends the event record to all its neighbors for which it has a gradient for this event. It continues to do so to each neighbor at the appropriate frequency, until the interest from that neighbor expires.

A node receiving an event record from its neighbors looks to see if it has any matching interests in its cache. If not, the record is dropped. Each node also maintains a data cache recording recently seen event records and other data items. The current event record is also dropped if it appears in that data cache (indicating that the same information has already arrived via a different path)—thus preventing data forwarding loops. Otherwise the event record is added to the cache and is also forwarded to the relevant node neighbors, as indicated by the matching entries in the interest cache. In this fashion, the event record is propagated back toward the requesting sink. Note that both interest and data propagation are done through purely local operations and the source and sink never directly know about each other. This indirect way in which interests and data propagate and meet allows directed diffusion to quickly adapt to changes in the network topology, to phenomena of interest that move across the sensor field, and so on.

In our example, event records from the source node will start flowing back to the sink along multiple routes. This means that eventually the sink may be receiving the same data from multiple neighbors, albeit only at the low event rate initially requested (see Figure 3.10). At this point, the sink can select to *reinforce* certain gradients and weaken or eliminate others. For example, the first neighbor to report matching data back to the sink is likely to be on the minimum delay path to the source. The sink can reinforce that path by resending the same interest to that neighbor, but with a higher event rate request. The node receiving the reinforced interest will in turn need to send a reinforced interest to one of its neighbors, which can again be chosen to be the first one to receive data on the matching interest. In this fashion, an empirically low delay path is gradually reinforced and becomes the dominant path of getting data from the source to the sink (again see Figure 3.10). Since all interests have an expiration

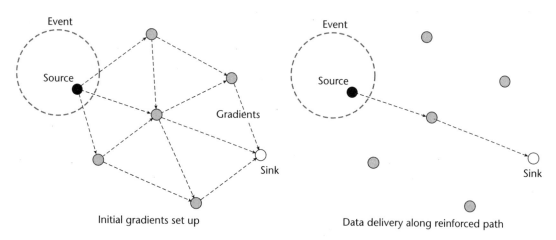

Figure 3.10 Directed diffusion sets up gradients for information delivery from the source to the sink (left), giving rise to multiple delivery paths. Reinforcement eventually redirects most of the information along the best path (right) (adapted from [104]).

duration, negative reinforcement can be applied by simply failing to renew interest requests along unproductive paths. The exact ways in which positive and negative reinforcements are to be applied can be subtle. The protocol has to balance efficiency in a stable setting (using only the best path available) against reactivity to changing conditions, which requires that data on the quality of multiple paths should always be available.

Even though in our simple scenario there was only a single source and sink, directed diffusion works equally well for shuttling information between multiple sources and sinks, both in a unicast and multicast manner. Note also that while our source/sink matching scenario was of the data "pull" type (all communications happened because of the sink's request), directed diffusion is sufficiently general to allow a "push" style of communication as well, letting sensor nodes trigger event propagation when they detect something that they believe might be of interest to potential sinks. Aggregation of data flowing from multiple sources to one sink can also be easily supported within directed diffusion.

In summary, directed diffusion provides a general-purpose communication mechanism for sensor networks. Unlike traditional networks that aim to provide point-to-point communication paths, directed diffusion is data-centric in its network view and performs all routing decisions through local, neighbor-to-neighbor interactions. It provides a reactive routing technique, discovering routes between information sources and sinks as needed. However, because of the data-centric view, data or data requests are propagated as part of the route discovery process. Once appropriate paths are discovered, the gradient mechanism can be exploited to concentrate traffic along the best paths, while always keeping enough of a peripheral view on changes in the network to allow for quick adaptation to new topologies or new demands. Experiments by the directed diffusion authors [104] show that it is also a highly energy-efficient protocol even when compared against protocols that use precomputed routes to link sources and sinks (omniscient shortest-path multicast trees). This is because directed diffusion can effectively suppress duplicate events and perform in-network information aggregation.

3.5.2 Rumor Routing

Directed diffusion resorts to initial flooding of the network in order to discover good paths between information sources and information sinks. But there are many situations in which, because the amount of data to be exchanged is small, the quality of the paths does not matter so much. In such situations, an attractive alternative is the technique of *rumor routing*, presented in [24]. Conceptually, in order to get sources and sinks to meet each other, we must spread information from each to regions of the sensor field, so that the two growing regions eventually intersect. In the previous section we discussed how directed diffusion spreads the interests generated at sinks in a two-dimensional fashion around each sink, while information sources do not distribute their information. Source and sinks get coupled only when the interest wave out of a sink reaches a corresponding source. As mentioned, one can also do this by propagating information from the sources, until they arrive at a matching sink. An interesting intermediate alternative is to spread information out of both sources and sinks, but in a *one-dimensional fashion*, effectively following a curve out of each—so as to reduce energy usage. When the curve emanating from a source meets the curve emanating from a matching sink, a path has been established between the two.

This is the central idea of rumor routing, which is ideally suited to using the routing on a curve protocol discussed in Section 3.4.2. In the simplest possible setting, the source information and sink interest trajectories are each a random straight ray emanating from their respective origin. It is convenient to think of an *agent* emanating out of each and propagating the data or interest. As the agents move, the corresponding data or interest is stored in all nodes the node has passed through. When there are multiple sources and sinks, it makes sense to merge information between different interest requests when they encounter each other, and also to merge information from different data propagating paths when they meet each other. So, for example, an interest agent passing though a node through which another interest agent has already passed can take that node's previously stored interest(s) and propagate them on its path, along with its own interest. Up to the limit of what can fit in a packet, this makes it more likely that matching data and interests will meet each other sooner, without adding traffic load to the network.

Related to rumor routing is the *ACQUIRE* query answering mechanism described in [198].[3] (See Chapter 6 for more information on the database aspects of sensor networks.) The ACQUIRE approach is most appropriate for situations when we need to process one-shot complex queries, whose answers depend on information obtained in several nodes of the network. For instance, we may want to count the number of horses detected by nodes at locations where the temperature is above some threshold. Because the query is not a continuous query (we are asking for this information just once), the gradient mechanisms of directed diffusion carry too much overhead. On the other hand, since there is spatial coherence in the parts of the network containing relevant information, rumor routing is also not the most efficient approach. The idea of the ACQUIRE system is to elect a leader node whose goal is to compute the answer to the query—initially this is the same node making the query. This node in turn enlists a number of its neighbors for help, using a tunable parameter describing the neighborhood size. After the leader has collected data from its neighbors, it selects one of them to be the new leader and the process repeats. The new leader can be selected either randomly or based on local information about the most promising direction in which to advance. The query is gradually resolved as more and more nodes are queried and their data aggregated, until a leader node decides that it has sufficient information to fully answer the query. At that point, the answer can be sent back to the originating query node, either by traversing the leader path backward or by direct geographic forwarding. The neighborhood size parameter is application dependent and trades off the cost per leader hand-off versus the number of such hand-offs. This is also the idea behind the information-driven sensor querying (IDSQ) [25]. The case of query using IDSQ will be discussed in Section 5.4.

---

3 *ACQUIRE* stands for ACtive QUery forwarding In sensoR nEtworks.

3.5.3 Geographic Hash Tables

There are many situations in which it is useful to view a sensor network as a distributed database storing observations and readings from the sensors for possible later retrieval by queries injected anywhere on the network. For example, instead of actively tracking horses and requesting a stream of closely spaced readings from the sensors near where horses currently are, we may simply want to know in what areas of the network horses were sensed over the past several days, assuming sensor nodes perform background readings at a low data rate. In such settings, since the data is sparse, the quality of the paths is less important. What is essential is to get to the observations about horses and the queries about horses to find each other in the network.

The *geographic hash table* (GHT) technique proposed in [191] accomplishes this by using sensor reading attributes to hash information to specific geographic locations in the network. Information records meeting those attributes are stored in nodes close to the hashed location. Queries about records having these attributes are routed to these nodes as well. Thus, in our example, nodes in the vicinity of the hash location become the "meeting point" between information providers and information seekers about horse sightings. Since the hash function spreads records around the network based on their attributes, this scheme evens out the load on the network by having each node participate in storing information as well as in routing data and queries to their destination. Information is stored locally but in a redundant fashion that allows recovery when nodes fail or move, thus making GHT a robust protocol.

One of the most elegant aspects of GHT is the way that it exploits features of the GPSR routing protocol (discussed in Section 3.4.2) to accomplish its goals. Consider trying to store a record r contained in a packet with information about a horse sighting. The hash function, given the attribute values type = animal and instance = horse, generates a location q in the plane (within the sensor field). In general there will not be a sensor node at exactly that location. The packet with r travels from its originating node toward its destination location using GPSR. The destination of the packet can be viewed as a virtual node q disconnected from the network. As we know, in such a situation GPSR will terminate in perimeter mode, with the packet with r going once in a cycle, around the face of the planar graph used by GPSR containing the point q. The initial node v through which the packet enters that cycle is the initial *home node* for the record r, and the cycle of nodes visited forms the *home perimeter* of the record r. GHT will store r redundantly at all nodes on its home perimeter. If the network is static, the *perimeter refresh protocol* part of GHT (described later) will guarantee that the home node moves around the perimeter cycle and stabilizes at the cycle node closest to the location q (see Figure 3.11). A node requesting information whose attributes hash to the same location q would again use GPSR and end up on the home perimeter of the record r, thus finding r.

As already mentioned, GHT uses a perimeter refresh protocol to accomplish replication of attribute-value pairs, ensure their consistent placement in the appropriate home node, and recover from node or link failures. The home node for a record r generates *refresh packets* periodically, addressed to the destination location q. With GPSR these packets go around the home perimeter. When a refresh packet arrives at a node w, if that node is closer to location q than the original home node v, then w declares itself to be the new home node, kills v's refresh packet, and initiates its own refresh packet. Otherwise, it passes v's packet on around the cycle.

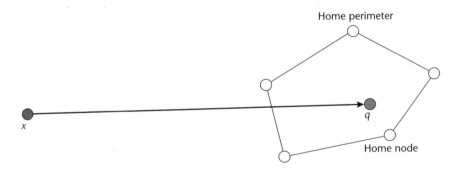

Figure 3.11   A packet is routed using GPSR to a location where no node exists. The home perimeter defines a ring of nodes around that location, and one of these is selected to be the home node for the packet.

In both cases, any additional records w has that match that particular attribute-value pair will be appended to the packet. In this way, all records hashing to locations inside a face in GPSR's planar graph will eventually be distributed to all nodes forming the cycle surrounding that face. Note that GPSR can detect when a packet has made a complete tour around the cycle, as in perimeter mode GPSR stores the identity of the first traversed edge from a cycle into the packet.

A nice feature of the perimeter refresh protocol is that it helps recover from node failures. This is accomplished in part by having in each node a takeover timer for each packet received and not originated at the node. A takeover timer rings if no refresh packet has been received from the home node after a duration equal to a few times what it takes to go around the home perimeter. At that point, the node holding the timer declares itself to be the home node for the record in question and initiates its own refresh packet. To prevent hot spots—that is, excessive access to information at one place—GHT also employs structured replication which mirrors the information stored at one node at others across the network, following the hierarchical scheme used for a geographic location service in [134].

3.6 Summary

In this chapter, we have surveyed a number of important ideas for networking sensor networks. These include specialized MAC protocols, routing algorithms based on geographic information, energy awareness and load balancing, as well as attribute-based distributed matching for mating information seekers and providers. The material in this chapter forms the foundation for several other topics we will discuss later in this book. A number of key themes are worth special mention:

- Network nodes operate in a collaborative manner, usually serving a single (or few) operational task(s). Thus issues of fair utilization of networking resources at the node level are less important than just accomplishing the overall goal. For example, in Chapter 5 we will investigate how routing and information aggregation can be jointly optimized in support of a localization task.

- Nodes also are ephemeral—they can be destroyed, be asleep, or die from malfunction or battery depletion. At the same time, nodes can be expendable and redundant, as information sensed by nearby nodes may be similar. What is important to the sensor network is the information the nodes contain, not the nodes themselves.

- Rapid changes in the topology of the network, due to status changes on nodes or links, argue for reactive routing protocols that discover paths as needed. This is why geographic routing is so important for sensor networks.

- Preservation of energy is of paramount concern. This affects all layers of the network stack, from medium access control, to routing, to data aggregation and in-network processing that may be application-specific. The death of a node can have consequences far beyond the node itself, if, for example, that node is critical to the connectivity of the network.

- All these considerations make networking for sensor networks data-centric as opposed to node-centric. The distinction between address and data in a packet is blurred, and nodes can make routing decisions based on the contents of a packet, as in directed diffusion and other protocols.

Even though routing is one of the most developed areas in sensor networks, additional research is needed across all topics touched upon in this chapter, from low-level MAC layer issues to high-level schemes that allow information suppliers and seekers to locate each other in the network, with or without an associated distributed storage mechanism. Since in-network processing is key to efficient use of network resources for many applications, we are likely to see the development of new programming abstractions that combine data processing tasks, such as data aggregation, with networking. As these abstractions become clarified and codified, they are likely to percolate down to even the lower layers of the network stack.

4

Infrastructure Establishment

When a sensor network is first activated, various tasks must be performed to establish the necessary infrastructure that will allow useful collaborative work to be performed. In particular, each node must discover which other nodes it can directly communicate with, and its radio power must be set appropriately to ensure adequate connectivity. Nodes near one another may wish to organize themselves into clusters, so that sensing redundancy can be avoided and scarce resources, such as radio frequency, may be reused across non-overlapping clusters. Furthermore, for most high-level collaborative network functions, the sensor nodes must be placed in a common temporal and spatial framework. This is essential in order to be able to reason about events sensed at various locations on different occasions. In this chapter, we survey some common techniques used to establish such infrastructure: topology control, clustering, time synchronization, and localization for the network nodes. We also describe how to implement a location service that can map node IDs to node locations, as needed for geographic routing.

4.1 Topology Control

A sensor network node that first wakes up executes a protocol to discover which other nodes it can communicate with (bidirectionally). This set of neighbors is determined by the radio power of the nodes as well as the local topography and other conditions that may degrade radio links. Unlike wired networks, however, nodes in a wireless sensor network can change the topology of the network by choosing

103

to broadcast at less than their maximum possible power. This can be advantageous in situations where there is dense node deployment, as radio power is the main drain on a node's batteries. The problem of *topology control* for a sensor network is how to set the radio range for each node so as to minimize energy usage, while still ensuring that the communication graph of the nodes remains connected and satisfies other desirable communication properties.

Although in principle the transmitting range of each node can be set independently, let us first examine the simpler case where all nodes must use the same transmission range: inexpensive radio transmitters, for example, may not allow the range to be adjusted. We also ignore all effects of interference or multipath, so that any pair of nodes within range of each other can communicate. This *homogeneous* topology control setting defines the *critical transmitting range* (CTR) problem: compute the minimum common transmitting range r such that the network is connected.

The solution to the CTR problem depends on information about the physical placement of the nodes—another of the infrastructure establishment tasks. If the node locations are known a priori, or determined using the techniques described later in this chapter (see Section 4.4), then the CTR problem has a simple answer: the critical transmitting range is the length of the longest edge of the minimum Euclidean spanning tree (MST) connecting the nodes. This easily follows from the property that the MST contains the shortest edge across any partition of the nodes. The MST can be computed in a distributed fashion, using one of several such algorithms in the literature [71].

The CTR problem has also been studied in a probabilistic context, where the node positions are not known but their locations come from a known distribution. The problem now becomes to estimate the range r that guarantees network connectivity with high probability (probability that tends to 1 as the number n of nodes grows to infinity). Such results are useful in settings where the node capabilities and mode of deployment prevent accurate localization. The probabilistic theory best suited to the analysis of CTR is the theory of *geometric random graphs* (GRG) [53]. In the GRG setting, n points are distributed into a region according to some distribution, and then some aspect of the node placement is investigated. For example, in [174] it is shown that if n points are randomly and uniformly distributed in the unit square, then the critical transmission range is, with high probability,

$$r = c\sqrt{\frac{\log n}{n}},$$

for some constant $c > 0$. Such asymptotic results can help a node designer set the transmission range in advance, so that after deployment the network will be connected with high probability.

Most situations, however, can benefit from allowing nodes different transmission ranges. Intuitively speaking, one should choose short ranges in areas of high node density and long ranges in regions of low density. If nodes can have different transmission ranges, then the goal becomes to minimize

$$\sum_{1}^{n} r_i^{\alpha},$$

where r_i denotes the range assigned to node i and α is the exponent describing the power consumption law for the system. This is the *range assignment* problem. Unfortunately, this problem has been shown to be NP-complete for dimensions two and above [44]. A factor 2 approximation can be computed by first building an MST on the nodes, where the weight of the edge connecting nodes i and j is $\delta^{\alpha}(i, j)$ [here $\delta(i, j)$ denotes the Euclidean distance from i to j]. The range r_i for node i is then set to be the maximum of $\delta(i, j)$ over all nodes j which are neighbors of i in the MST [116].

The homogeneous or nonhomogeneous MST-based algorithms can be expensive to implement on typical sensor nodes. Several protocols have been proposed that attempt to directly solve the CTR problem in a distributed way. For example, the COMPOW protocol of [166] computes routing tables for each node at different power levels; a node selects the minimum transmit power so that its routing table contains all other nodes. Recent work has also focused on topology control protocols that are lightweight and can work with weaker information than full knowledge of the node positions. For an excellent survey of these protocols and the entire topology control area, the reader is referred to [201].

4.2 Clustering

The nodes in a sensor network often need to organize themselves into clusters. Clustering allows hierarchical structures to be built on the nodes and enables more efficient use of scarce resources, such as frequency spectrum, bandwidth, and power. For example, if the cluster size corresponds roughly with the direct communication range of the nodes, much simpler protocols can be used for routing and broadcasting within a cluster; furthermore, the same time or frequency division multiplexing can be reused across nonoverlapping clusters. Clustering also allows the health of the network to be monitored and misbehaving nodes to be identified, as some nodes in a cluster can play watchdog roles over other nodes [153]. Finally, networks can be comprised of mixtures of nodes, including some that are more powerful or have special capabilities, such as increased communication range, GPS, and the like. These more capable nodes can naturally play the role of *cluster-heads*.

In many cases, however, the nodes are identical and their common communication range is a natural cluster size. There exist several distributed protocols for cluster-head election in this setting, mostly based on node unique identifiers (UIDs)—these can be any unique IDs assigned to the nodes initially [15, 80, 76]. In some variants, a node declares itself a cluster-head if it has a higher ID than all its "uncovered" neighbors—neighbors that have not been already claimed by another cluster-head. In others, each node nominates as a cluster-head the highest ID node it can communicate with (including itself). Nominated nodes then form clusters with their nominators.

Nodes that can communicate with two or more cluster-heads may become *gateways*—nodes that aid in passing traffic from one cluster to another. In some applications, it may be useful to view the IDs as weights, indicating which nodes are to be favored in becoming cluster-heads.

Such simple protocols work well in practice, even though no theoretical guarantees can be proven about the quality of the clustering (the number of clusters obtained as compared with the minimum possible), even in a randomized setting (where we assume the node IDs are randomly distributed). A constant approximation bound can be shown if the leader election protocol is used hierarchically, with exponentially increasing node ranges, until the hardware provided maximum range is reached [76].

Clustering can be used to thin out parts of the network where an excessive number of nodes may be present. A simplified long-range communication network can be set up using only cluster-heads and gateways—all other nodes communicate via their cluster-head. Cluster-heads can be chosen to have a minimum separation comparable to the node communication range. This property ensures that each cluster-head has a bounded number of cluster-head neighbors and that the density of cluster-heads is bounded from above as already discussed in Section 3.4.1. Several local communication protocols can then become simpler, since in the inner loop of such protocols a node needs to query all of its neighbors for some information.

Additional research is needed into how to get all the benefits of clustering while distributing the load (and battery drain) of being a cluster-head evenly among all the nodes.

4.3 Time Synchronization

Since the nodes in a sensor network operate independently, their clocks may not be, or stay, synchronized with one another. This can cause difficulties when trying to integrate and interpret information sensed at different nodes. For example, if a moving car is detected at two different times along a road, before we can even tell in what direction the car is moving, we have to be able to meaningfully compare the detection times. And clearly we must be able to transform the two time readings into a common frame of reference before we can estimate the speed of the vehicle. Estimating time differences across nodes accurately is also important during node localization (see Section 4.4). For instance, many localization algorithms use ranging technologies to estimate internode distances; in these technologies, synchronization is needed for time-of-flight measurements that are then transformed into distances by multiplying with the medium propagation speed for the type of signal used (say, radio frequency or ultrasonic). Configuring a beam-forming array or setting a TDMA radio schedule are just two more examples of situations in which collaborative sensing requires the nodes involved to agree on a common time frame.

While in the wired world time synchronization protocols such as NTP [159, 160] have been widely and successfully used to achieve Coordinated Universal Time (UTC), these solutions do not transfer easily to the ad hoc wireless network setting. These wired protocols assume the existence of highly accurate master clocks on some network nodes (such as atomic clocks) and, more importantly, they also require that pairs of nodes in the protocol are constantly connected and that they experience consistent communication delays in their exchanges. For the case where there is variability in delays, but this variability can be described by a nice distribution, probabilistic clock synchronization methods have been proposed [42, 172]. Unfortunately, none of these assumptions is generally valid in sensor networks. No special master clocks are available, connections are ephemeral, and communication delays are inconsistent and unpredictable. Thus we must moderate our goals when it comes to synchronizing node clocks in sensor networks. In our setting, we may be satisfied with local as opposed to global synchronization, as collaborative processes often have a strongly local character. Furthermore, for many applications that involve temporal reasoning about the phenomena being sensed, only the time ordering of event detections matters and not the absolute time values.

4.3.1 Clocks and Communication Delays

Computer clocks are based on hardware oscillators which provide a local time for each sensor network node. At real time t the computer clock indicates time $C(t)$, which may or may not be the same as t. For a perfect hardware clock, the derivative $dC(t)/dt$ should be equal to 1. If this is not the case, we speak of clock *skew* (also called *drift*). The clock skew can actually change over time due to environmental conditions, such as temperature and humidity, but we will assume it stays bound close to 1, so that

$$1 - \rho \leq \frac{dC(t)}{dt} \leq 1 + \rho,$$

where ρ denotes the maximum skew. A typical value of ρ for today's hardware is 10^{-6}. Small fluctuations on the skew are usually modeled as random Gaussian noise. Note that, because of clock skew, even if the clocks of two nodes are synchronized at some point in time, they need not stay so in the future.

Even if no skew is present, the clocks of different nodes may disagree on what time "0" means. Time differences caused by the lack of a common time origin are referred to as clock *phase* differences (or clock *bias*).

For nodes to be able to synchronize, they must have for a period a communication channel where message delays can be reliably estimated. The latency in channel can be decomposed into four components [63]:

- *Send time:* This is the time taken by the sender to construct the message, including delays introduced by operating system calls, context switching, and data access to the network interface.

- *Access time:* This is the delay incurred while waiting for access to the transmission channel due to contention, collisions, and the like. The details of that are very MAC-specific.

- *Propagation time:* This is the time for the message to travel across the channel to the destination node. It can be highly variable, from negligible for single-hop wireless transmission to very long in multihop wide-area transmissions.

- *Receive time:* This is the time for the network interface on the receiver side to get the message and notify the host of its arrival. This delay can be kept small by time-stamping the incoming packet inside the network driver's interrupt handler.

If there is no clock skew and the (unknown) delay D in the channel between two nodes is constant and perfectly symmetric, then two nodes i and j can estimate their phase difference d with three message exchanges, as follows. See Figure 4.1.

- Node i reads its local clock with time value t_1 and sends this in a packet to node j.

- Node j records the time t_2 according to its own clock when the packet was received. We must have $t_2 = t_1 + D + d$.

- Node j, at time t_3, sends a packet back to i containing t_1, t_2, and t_3.

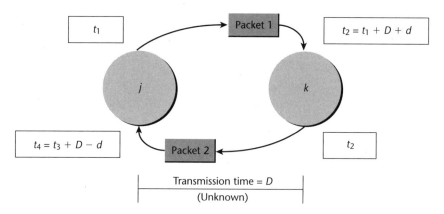

Figure 4.1 Clock phase difference estimation, using three message exchanges (adapted from [104]).

- Node i receives this packet at time t_4. We must have $t_4 = t_3 + D - d$. Therefore, node i can eliminate D from the above two equations and compute $d = (t_2 - t_1 - t_4 + t_3)/2$.

- Finally, node i sends the computed phase difference d back to node j.

Time synchronization can then be propagated across the network by using a spanning tree favoring direct connections with reliable delays [220]. In the presence of clock skew, however, frequent resynchronizations may be required. Furthermore, such ideal conditions on delays are hardly ever true in a sensor network.

In the presence of clock skew and phase differences, as well as variability in transmission delays, we present two methods that accomplish a more modest form of clock synchronization for sensor networks. In both cases, the approach provides mappings from clock readings made on one node to the clocks of other nodes, without an attempt to synchronize multiple nodes to a common global clock, as in traditional networks. For protocols that aim to synchronize across a wide area network, the number of messages needed represents a serious overhead and power drain and thus must be kept to a minimum.

4.3.2   Interval Methods

As we mentioned, in many situations involving temporal reasoning, the temporal ordering of events matters much more than the exact times when events occurred. In such situations, interval methods provide a lightweight protocol that can be used to move clock readings around the network and perform temporal comparisons [194].

Suppose that event E occurs at real time $t(E)$ and is sensed by some node i and given a time stamp $S_i(t)$, according to the local clock of node i. Suppose also that node i's clock runs with a known maximum skew of ρ_i. The key idea of the protocol is to focus on time intervals between events and transform such time differences from the time framework of one node to that of another by estimating communication delays and applying interval methods. Since what is being transformed is the difference between two local times, phase differences do not matter as they cancel out—but clock skew and network latency have to be dealt with. We will call intervals between events *durations*.

In the simplest setting, if node 1 with maximum clock skew ρ_1 wishes to transform a local duration ΔC_1 into the time framework of node 2 with maximum clock skew ρ_2, we can proceed as follows. If the real time duration is Δt, then we must have

$$1 - \rho_i \leq \frac{\Delta C_i}{\Delta t} \leq 1 + \rho_i,$$

for $i = 1, 2$. Thus the real time duration Δt is contained in the interval $[\Delta C_1/(1 + \rho_1), \Delta C_1/(1 - \rho_1)]$, and the duration according to the clock of node 2 satisfies

$$\Delta C_2 \subseteq \left[\Delta C_1 \frac{1 - \rho_2}{1 + \rho_1}, \Delta C_1 \frac{1 + \rho_2}{1 - \rho_1} \right].$$

Now suppose nodes 1 and 2 are neighbors and have a direct communication link between them. Node 1 has detected an event E and time-stamped it with time stamp $r_1 = S_1(E)$. This temporal event needs to be communicated and transformed into the temporal frame of node 2. We must estimate the communication delay between the nodes. Now, under most communication protocols, for every message M that node 1 sends to node 2, there is a return acknowledgment message A from node 2 to node 1. Node 1 can measure the duration d between transmitting M and receiving A and use that as an upper bound on the communication delay (the obvious lower bound is 0). However, it is node 2 that needs to know this information in order to update the time stamp generated by node 1. This seems to require two message exchanges between nodes 1 and 2: a message M_1 carrying r_1, and a subsequent message M_2 carrying d (along with the corresponding acknowledgments A_1 and A_2), thus effectively doubling the communication overhead (see Figure 4.2).

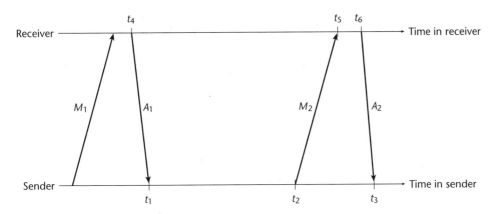

Figure 4.2 Transforming time stamps from the reference frame of one node to that of another (adapted from [194]).

Because communicating nodes typically exchange several messages, it makes sense to piggyback delay information on other content-carrying messages. Let us look again at Figure 4.2. Suppose M_1 was a message sent from node 1 to node 2 earlier, for other purposes. Now M_2 will be used to carry information about the time stamp $S_1(E)$. The idle time duration $\ell_1 = t_2 - t_1$ can be measured according to the local clock of node 1, as the time between receiving A_1 and transmitting M_2. The round-trip duration $p_1 = t_5 - t_4$ can be measured according to the local clock of node 2, as the time between transmitting A_1 and receiving M_2. If sender node 1 piggybacks the idle time duration ℓ_1 on M_2, then at time t_5, the receiver node 2 can estimate the communication delay d via the bounds

$$0 \leq d \leq p_1 - \ell_1 \frac{1 - \rho_2}{1 + \rho_1}.$$

Of course this assumes the two nodes had earlier communication in the recent past and that, if a node communicates with several others, it keeps track of its last communication to each of its neighbors.

Given this setting, time stamps can be propagated from node to node as follows. Let r_i and s_i denote, respectively, the times when node i receives and sends out the packet containing the time stamp (measured according to its local clock). Let ℓ_i and p_i denote the corresponding idle and round-trip times, as earlier (note that ℓ_i is measured in the clock of node $i - 1$). Then we can recursively maintain a valid interval guaranteed to contain the original time stamp according to the local clock of node i, as follows.

For node 1, the interval is

$$[r_1, r_1] = [S_1(E), S_1(E)].$$

For node 2, using the above reasoning, the interval is

$$\left[r_2 - (s_1 - r_1)\frac{1 + \rho_2}{1 - \rho_1} - \left(p_1 - \ell_1 \frac{1 - \rho_2}{1 + \rho_1} \right), \ r_2 - (s_1 - r_1)\frac{1 - \rho_2}{1 + \rho_1} \right].$$

And for the nth node in the transmission chain we get by iterating

$$\left[r_n - (1 + \rho_n) \sum_{i=1}^{n-1} \frac{s_1 - r_i + p_i}{1 - \rho_i} - p_{i-1} + (1 - \rho_n) \sum_{i=1}^{n-1} \frac{\ell_i}{1 + \rho_i}, \right.$$

$$\left. r_n - (1 - \rho_n) \sum_{i=1}^{n-1} \frac{s_i - r_i}{1 + \rho_i} \right].$$

Thus in the end, detection times at one node can be transformed to time intervals in the local time frame of another node through a sequence of one-hop communications. Comparison of time stamps is done through standard interval methods: if the corresponding intervals are disjoint, then a meaningful time-stamp comparison can be made, and otherwise not.

This time synchronization protocol has low overhead, scales well with network size, and can accommodate topology changes and short-lived connections. But, as with most interval methods, the intervals computed can get too large to be useful or to resolve most of the time comparisons needed. It should be clear that the size of intervals increases roughly linearly with the number of hops, as well as with the length of time a node holds a time stamp before retransmission (the $s_i - r_i$ term). Long holding times might arise because of temporary disconnections in the network. Furthermore, in a topology where multiple paths are possible between nodes that need to compare time stamps, the problem of which communication path to select for synchronization deserves further investigation.

4.3.3 Reference Broadcasts

Time comparisons are not sufficient for all applications and mappings from event times to time intervals may quickly become useless if long delays or multihop routes increase the interval sizes beyond reasonable limits. Note that even if there is no skew between the clocks of different nodes (say $\rho_i = 1$ for all nodes above), time intervals can still grow large because delay estimates can have large uncertainty.

The key idea of the *reference broadcast system* (RBS) of [63] is to use the broadcast nature of the wireless communication medium to reduce delays and delay uncertainty in the synchronization protocol. This is achieved by having the receiver nodes within the communication range of a broadcast message sent by a sender node synchronize with one another, rather than with the sender. This is accomplished by having the sender send a reference message to receivers who record its time of arrival each in their own time frame. The receivers then exchange this information among themselves. The advantage of doing this is that since all receivers receive the same message packet, they all share exactly the same send time and access time delays in the communication, and the propagation time is negligible anyway for single-hop wireless communication. Thus the only source of variability and nondeterminism in the time at which the receivers are notified of the receipt of this packet is the variability in the receive time among the different receiver nodes.

As we remarked earlier, the receive delay can be kept small by time-stamping the incoming packet in the network driver interrupt handler of the receiver, thus bypassing context switching and procedure call delays in the host. Furthermore, delays within the network interface card hardware need not affect the synchronization, as long as they are the same for all receiving nodes.

If we assume no clock skew, then the message exchange between the receivers can be used to estimate the relative phase differences, up to the uncertainties in receive times, which we have just argued can be kept small (mostly under 10 μsec in the experiments of [63] with the Berkeley motes). Receive times can be subject to random fluctuations because of environmental conditions; one way to mitigate the impact of these nondeterministic variations is to repeat the reference broadcast protocol a number of times and then average results in computing internode time offsets (relative phase differences). The group dispersion (i.e., the maximum offset between any pair of receiver nodes) can be significantly decreased in a large group of receiver nodes by such statistical methods (say, by a factor of 5, by repeating the reference broadcast 20 times [63]).

Of course, clock skew has to be dealt with as well. One way to estimate both clock skew and phase differences among pairs of receivers is to do a least-squares linear regression among all the pairwise measurements obtained for the two receivers in a sequence of reference broadcasts. This effectively provides an affine function mapping clock readings from the frame of one receiver to that of the other. A linear regression assumes that the clock skew is fixed, of course, but this is a reasonable assumption for a series of reference broadcasts closely spaced in time.

The RBS protocol as described up until now deals with synchronization only among nodes that are in range of the same sender. Multihop synchronization is also possible, by just composing the inter-receiver clock affine mappings in the right way. Consider the scenario in Figure 4.3.

In this figure, sender A can synchronize nodes 1, 2, 3, and 4, while sender B can synchronize nodes 4, 5, 6, and 7 (of course, the sender/receiver distinction is purely artificial—any node can be sender or receiver). The affine clock maps obtained by RBS between nodes 1 and 4 (through A), and nodes 4 and 7 (through B), for example, can be composed to provide the clock map between nodes 1 and 7. In general, we can imagine an RBS graph with an edge

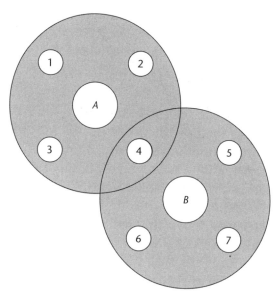

Figure 4.3 The restricted broadcast time synchronization protocol between nonneighboring nodes (adapted from [63]).

between any two nodes that can be reached in one hop from a common sender. The edge between these nodes can be given a weight corresponding to the uncertainty of the clock phase difference estimation between the two nodes. Then, whenever two nodes need to synchronize in the graph, a shortest path between them can be sought using standard shortest-path graph algorithms [46]. Such algorithms can be expensive to run in a distributed sensor network setting; an alternative is to use the RBS pairwise affine clock mappings for neighboring nodes to transform times in a packet to the local frame of the current receiver, as the packet travels through the network, much in the spirit of the interval methods described in Section 4.3.2.

4.4 Localization and Localization Services

An attractive feature of a sensor network is that it can provide information about the world that is highly localized in space and/or time.

For several sensor net applications, including target tracking and habitat monitoring, knowing the exact location where information was collected is critical. In fact, for almost all sensor net applications, the value of the information collected can be enhanced if the location of the sensors where readings were made is also available. With current technologies, most sensor nodes remain static. Thus one way to know the node positions is to have the network installer measure these locations during network deployment. This may not always feasible, however, especially in ad hoc deployment scenarios (such as dropping sensors from an aircraft), or in situations where the nodes may need to be moved for a variety of reasons (or move themselves, when that capability can be incorporated).

In this section, we describe techniques for *self-localization*—that is, methods that allow the nodes in a network to determine their geographic positions on their own as much as possible, during the network initialization process. We also describe *location service* algorithms—methods that allow other nodes to obtain the location of a desired node, after the initial phase in which each node discovers its own location. Such location services are important for geographic routing, location-aware query processing, and many other tasks in a sensor network.

Since the availability of GPS systems in 1993, it has been possible to build autonomous nodes that can localize themselves within a few meters' accuracy by listening to signals emitted by a number of satellites and assistive terrestrial transmitters. But even today GPS receivers can be expensive and difficult to incorporate into every sensor node for a number of practical reasons, including cost, power consumption, large form factors, and the like. Furthermore, GPS systems do not work indoors, or under dense foliage, or in other expectable conditions. Thus in a sensor network context, it is usually reasonable to assume that some nodes are equipped with GPS receivers, but most are not. The nodes that know their position are called *landmarks*. Other nodes localize themselves by reference to a certain number of landmarks, using various ranging or direction-of-arrival technologies that allow them to determine the distance or direction of landmarks.

4.4.1 Ranging Techniques

Ranging methods aim at estimating the distance of a receiver to a transmitter, by exploiting known signal propagation characteristics. For example, pairs of nodes in a sensor network whose radios are in communication range of each other can use *received signal strength* (RSS) techniques to estimate the RF signal strength at the receiver. If the source signal strength is known, along with the attenuation law for signal strength as a function of distance, then the receiver node can use RSS to estimate its distance from the sender. RF signal strength attenuation is not unlike the acoustic signal attenuation discussed earlier (Chapter 2)—they both drop off as an inverse power of the distance. Such a distance estimate, however, is usually not very accurate because RSS can vary substantially owing to fading, shadowing, and multipath effects. Variations in height between sender and receiver can also affect the measurement accuracy. Furthermore, radios in typical sensor nodes, because of cost considerations, do not come with well-calibrated components, so the source signal strength value may exhibit significant fluctuations. In general, localization to within a few meters is the best that can currently be attained with RSS methods.

A second way to estimate distance is to measure the time it takes for a signal to travel from sender to receiver; this can be multiplied by the signal propagation speed to yield distance. Such methods are called *time of arrival* (TOA) techniques and can use either RF or ultrasound signals. This requires that the sender and receiver are synchronized, and that the sender knows the exact time of transmission and sends that to the receiver. Since the exact time of transmission may be hard to know for the reasons discussed in the previous section, an alternative is to measure the *time difference of arrival* (TDOA) at two receivers, which then lets us estimate the difference in distances between the two receivers and the sender. Another issue is that signal propagation speed exhibits variability as a function of temperature or humidity as well (especially for ultrasound), and thus it is not realistic to assume it is constant across a large sensor field. Local pairs of beacons can be used to estimate local propagation speed. With proper calibration and the best current techniques, localization to within a few centimeters can be achieved, as reported in the experimental test-bed described in [204].

For more information on ranging techniques, the reader is referred to [81].

4.4.2 Range-Based Localization Algorithms

We now describe methods for localizing sensor network nodes with reference to nearby *landmarks*—we shall use the latter term to refer to other nodes that have already been localized. We confine our attention to distance measurements, obtained using one of the ranging techniques described earlier.

The position of a node in the plane is determined by two parameters: its *x* and *y* coordinates. Therefore, at least two constraints are necessary to localize a node. A distance measurement with respect to a landmark places the node on a circle centered at the landmark whose radius is the measured distance (in the TDOA case, a difference of distances to two landmarks places the node on a hyperbola with the landmarks as foci). Since quadratic curves in the plane can have multiple intersections, in general a third distance measurement is necessary in order to completely localize a node (see Figure 4.4).

In fact, most measurements have error, so the node in question is only localized to within a band around the measured distance circle. For this reason, it may be advantageous to use redundant measurements and least-squares techniques to improve the estimation accuracy. With additional measurements, in TOA methods, the propagation speed can be estimated locally as well, which will yield improved localization accuracy, as mentioned earlier.

We now describe this operation, called *atomic multilateration*, in some detail [204]. The analysis is similar to the collaborative source localization discussed in Section 2.2.2. Suppose we number the node whose location we seek as node 0 and the available landmark nodes as $1, 2, \ldots, n$. Let the position of node i be (x_i, y_i) and its measured

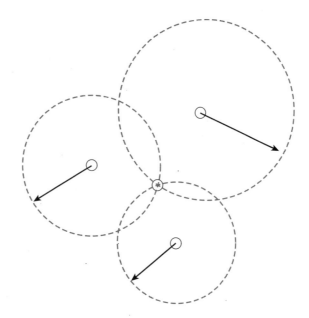

Figure 4.4 Localizing a node using three distance measurements.

time of arrival at node 0 be t_i (for $1 \leq i \leq n$). If s denotes the local signal propagation speed, then we have for each i, $1 \leq i \leq n$

$$\sqrt{(x_i - x_0)^2 + (y_i - y_0)^2} + \epsilon_i(x_0, y_0, s) = st_i,$$

where ϵ_i indicates the error in the ith measurement due to noise and other factors. We can also give each measurement a relative weight α_i, indicating how much confidence we want to place in it. Our goal then is to estimate x_0, y_0, and s so as to minimize the weighted total squared error

$$E(x_0, y_0, s) = \sum_{i=1}^{n} \alpha_i^2 \epsilon_i^2(x_0, y_0, s).$$

For simplicity, we assume below that we have set $\alpha_i = 1$ for all i.

We can linearize the above system of n constraints by squaring and subtracting the equation for measurement 1 from that of the others, thus obtaining $n - 1$ linear equations of the form (the $x_0^2 + y_0^2$ terms cancel):

$$2x_0(x_i - x_1) + 2y_0(y_i - y_1) + s^2(t_i^2 - t_1^2) = -x_i^2 - y_i^2 + x_1^2 + y_1^2.$$

Here we treat s^2 as a new variable. We can write this in matrix form as $v = uA$, where

$$u = \begin{bmatrix} x_0 \\ y_0 \\ s^2 \end{bmatrix},$$

$$A = \begin{bmatrix} 2(x_2 - x_1) & 2(y_2 - y_1) & t_2^2 - t_1^2 \\ 2(x_3 - x_1) & 2(y_3 - y_1) & t_3^2 - t_1^2 \\ \cdot & \cdot & \cdot \\ \cdot & \cdot & \cdot \\ \cdot & \cdot & \cdot \\ 2(x_n - x_1) & 2(y_n - y_1) & t_n^2 - t_1^2 \end{bmatrix},$$

and

$$v = \begin{bmatrix} -x_2^2 - y_2^2 + x_1^2 + y_1^2 \\ -x_3^2 - y_3^2 + x_1^2 + y_1^2 \\ \cdot \\ \cdot \\ \cdot \\ -x_n^2 - y_n^2 + x_1^2 + y_1^2 \end{bmatrix}.$$

This leads to a least-squares problem whenever $n \geq 4$. A unique solution is possible whenever the landmarks are not collinear (in that case, A is clearly singular and there are two solutions symmetric about the line containing the landmarks). The solution to the system is given by $u = [(A^T A)^{-1} A^T]v$, which is also the maximum likelihood solution under the assumption of Gaussian measurement noise. Many other approaches to atomic multilateration have also been studied. See Chapter 2 of [111] and the many references therein.

Of course, it is unlikely that in a sensor network the density of nodes with GPS will be so high that every node will have three or four noncollinear landmarks within range initially. In some cases, the above atomic multilateration algorithm can be applied in an iterative fashion: those nodes with sufficient landmarks in range get themselves localized and in turn become landmarks for other nodes, and so on. See Figure 4.5 for an example. Initially nodes 1, 2, 8, 9, and 15 are landmarks. Node 10 can estimate its location using landmarks 8, 9, and 15. After that, node 7 can estimate its position using nodes 8, 9, and 10 as landmarks. This process terminates when no unlocalized node has enough landmarks to determine its position. Localization errors propagate in this process, and it is usually wise to always first localize the nodes with the most landmarks, as their location may carry the greatest certainty. But whether this iterative

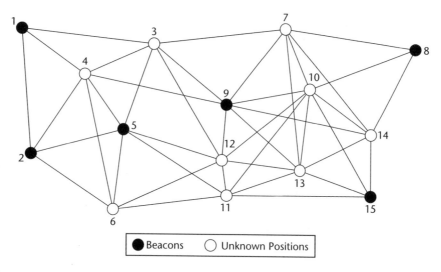

Figure 4.5 Iterative multilateration successively localizes more and more nodes (adapted from [203]).

algorithm is executed centrally or in a distributed fashion, there will be node configurations where this process will be unable to localize all the nodes.

In the general case, iterative multilateration fails. What we have to go by is the positions of the original landmark nodes with GPS, plus various distance estimates between node pairs. In principle, if we have enough constraints and no degeneracies are present, we can write this as a large system of nonlinear equations that can be solved for the unknown node positions. However, solving such global algebraic systems is expensive, and the solution must be computed in a centralized fashion. To get around these difficulties, a method titled *collaborative multilateration* is described in [203] which admits of a reasonable distributed implementation.

Collaborative multilateration proceeds by computing substructures of the network called *collaborative subtrees*—these are subgraphs of the full network graph in which there are enough constraints to make the localization problem sufficiently overdetermined that error accumulation is avoided. For each collaborative subtree, initial node positions are estimated and then refined using a Kalman filter technique, sequentially incorporating each successive measurement. Nodes that are not part of any collaborative subtree have their positions estimated only at the end, using distances to nodes already placed.

Collaborative subtrees are built using the notion of *tentatively unique* nodes. A node is called tentatively unique during a multilateration sequence if its position can be uniquely determined, assuming the positions of the other nodes used as references are unique. A recursive algorithm is used to traverse the network from a given node and accumulate nodes that can form a collaborative subtree. Initial position estimates are obtained by computing bounding boxes for each node based on its distance constraints and the known landmark locations, as projected on the x- and y-axes. The initial location for a node is the center of each bounding box. The position refinement, if done centrally, is best performed by a Kalman filter implementation, where the node positions are reestimated as the distance constraints are introduced one after the other. For further details, as well as a description of the distributed implementation, the reader is referred to [203].

An analysis of the errors in these multilateration processes and their dependence on node density is given in [205].

4.4.3   Other Localization Algorithms

In settings where RSS and other ranging technologies cannot be used directly to estimate distances, there are a number of alternatives. In every sensor network, each node knows what other nodes it can talk to directly—its one-hop neighbors. If the sensor nodes are densely and uniformly deployed, then we can use hop counts to landmarks as a substitute to physical distance estimates. In this setting, each landmark floods the network with a broadcast message whose hop count is incremented as it is passed from node to node. The hop count in the message from a landmark that first reaches a node is the hop distance of that node to the landmark (standard graph-based breadth-first search). In order to transform hop counts into approximate distances, the system must estimate the average distance corresponding to a hop. This can be done either by using inter-landmark distances that are known in both hop and Euclidean terms [169], or by using prior information about the size of the area where the nodes are deployed and their density [165]. Once a node has approximate distances to at least three landmarks this way, then the previous uni- and multilateration techniques can be used.

Note that the presence or absence of a radio link between two nodes is a quantized form of received signal strength measurement. The general effect of quantizing RSS on the quality of sensor localization is discussed in [176]. If the radio range is known, the presence of a radio link can also be expressed as a convex constraint on the positions of the nodes, and convex programming techniques have been suggested for addressing the localization problem [56].

Another interesting option is to use RSS only for distance comparisons to landmarks and not for distance estimation. In the neighborhood of a node, a slight displacement that results in increased RSS from a landmark can be taken to indicate the node has moved closer to the landmark; correspondingly, reduced RSS from a landmark can be taken to indicate the node has moved farther from the landmark. Even if sensor nodes cannot move, they can interrogate their neighbors for their RSS and thus can make inferences about relative distances to landmarks. This becomes interesting for localization because of the following observation: if a node is contained in a triangle defined by three landmarks it can hear, then no matter how it moves differentially, the RSS from at least one of the three landmarks has to increase. On the other hand, if the node is outside the triangle defined by the three landmarks, then there will be some direction of motion so that RSS from all three landmarks decreases. Again, in the absence of motion, a node can perform an approximate version of this test by interrogating all of its one-hop neighbors.

In [93] the authors suggest using this *Approximate Point in Triangle* (APIT) test for range-free localization. For any triplet of landmarks that a node can hear, if the node passes the APIT test (i.e., for each one of its neighbors at least some landmark sounds stronger) with respect to these landmarks, then the node is declared to be in the triangle defined by the landmarks. Thus the node can be localized to lie at the intersection of all landmark triangles that are known to contain it. See Figure 4.6 for an example. It should be noted again that both false positives and false negatives are possible with APIT. A node

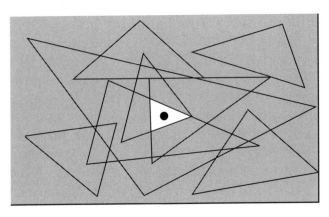

Figure 4.6 Node localization using multiple triangle containment tests (adapted from [93]).

may be outside a landmark triangle but simply fail to have a neighbor in the direction in which all three landmarks have reduced RSS (false positive). Or a node may be inside such a triangle but have a neighbor who is outside and for which all three landmarks have reduced RSS (false negative). Experiments reported in [93] indicate that APIT failures happen less than 14 percent of the time. The same paper argues that the accuracy of localization provided by k landmarks is $O(1/k^2)$.

4.4.4 Location Services

Some sensor network tasks can be accomplished by routing information to geographic regions of the network, as defined by the task demands. Protocols for accomplishing this were described in Chapter 3. But in other tasks, we may want to address messages to specific nodes whose geographic location is not known to us. We may know these nodes via their UID, or via some problem-specific attributes that uniquely identify them. For example, we may want to address the leader node of a node cluster tracking a particular target, as in Chapter 5. Thus it is not enough for each node to know only its own location. For geographic forwarding, energy-aware routing, and many other sensor net tasks, it is important to have a *location service*, a mechanism for mapping from node IDs of some sort to node locations. Note that even if the actual nodes in a sensor network are static, virtual node IDs, such as the tracking cluster leader mentioned earlier, can be mobile because of hand-offs from node to node (as the target moves).

Providing a location service is an interesting problem, because the obvious solutions of having a central repository that correlates IDs to locations, or having a copy of that at every node, suffer from significant drawbacks. The former has a single point of failure and can cause communication delays unrelated to the distance between the sender and receiver; the latter requires space on every node and is extremely expensive to update in case some (virtual or real) nodes move, or are inserted or deleted.

What we would like to have is a distributed location service that is robust to single node failures, spreads the load evenly across the network, and also has nice locality properties, in the sense that a node wishing to determine the location of another nearby node can do so at a cost that is sensitive to the distance of the receiver. Such a location service is in the *Grid* system presented in [134], which we describe below. We present this service using node UIDs as the node "names"; readers should keep in mind that any virtual name can be used equally well.

The Grid system uses a distributed location service termed GLS (Geographic Location Service). The service is implemented using the sensor nodes themselves as location servers. The key idea of GLS is to distribute the load so that each network node acts as a location server for a relatively small number of other nodes, most in its neighborhood. Each node u needs to inform the location servers that will handle u of its position and, symmetrically, each node making a location query for a given node v must be able to easily locate one of the servers that know about v. GLS accomplishes this by organizing the nodes into a hierarchical structure according to a spatial quad-tree, plus exploiting the ID assigned to each node. The node ID can be any unique name associated with a node, such as its MAC address—GLS only needs the property that the IDs are distinct and come from an ordered set. In a distributed manner, then, GLS provides a mapping from node IDs to node locations. Grid is also capable of handling node mobility and continuously updating the location server information as nodes move—which is useful when we want to use this facility for locating nodes based on virtual IDs that can hop around the network (emulating node mobility).

So which are the location servers that will be recruited to hold the location of a particular node? Let us assume that the leaf tiles of the quad-tree are small enough so that all nodes in a tile are in communication range of one another—thus a node B can behave as a location server for itself toward the other nodes within that tile, trivially. Let us say that leaf tiles are at level 1. The hierarchical idea of GLS is to first select a location server for B in each of the three sibling leaf tiles to the leaf tile containing B. Together these level-1 tiles form a tile at level 2 of the quad-tree—this is the unique level-2 tile containing B. Again B recruits location servers for itself in the three siblings of that level-2 tile, and so on, going up the tree. If we have n nodes that are reasonably uniformly distributed in a field, then the depth (maximum level) of the quad-tree will be $O(\log n)$ and the total number of servers recruited for B will also be $O(\log n)$. For example, see Figure 4.7, where numbers indicate node IDs. Among all nodes in a chosen tile, the one selected to be the location server for B is always the node "closest to B," defined to be *the node with the least ID greater than B*. The ID space is taken to wrap around, so that if there are no nodes greater than B in the tile, then we select the node with the smallest ID less than B. The circled nodes in the figure are the location servers selected.

In Figure 4.7 there is only one node per atomic quad-tree tile. For each such tile, Figure 4.8 shows in small size numbers the list of all nodes that have recruited the node in that tile as their location server. For random IDs, it can be shown that the expected length of that list is also $O(\log n)$.

But how does node B determine which is the appropriate location server node for it in a given tile? Node B simply used geographic forwarding to send a packet with its current location to the tile in question. The first node L in that tile to receive the packet initiates a search for the least ID node greater than B in that tile, who will be recruited as the location server for B. This latter search is part of the exact search used when a node wishes to locate a location server for B—a process we now describe.

A node A wishing to communicate with node B whose ID it knows sends a request using geographic forwarding to the node C with the least ID greater than or equal to B for which A has location information. If C knows the location of B, then it can use geographic forwarding to forward the requested communication to B directly, and then B can reply to A directly as well, since the location of A can be part of the initial message. If C does not know the location of B, then C repeats the above process, passing the message to the node D with the least ID greater than or equal to B for which C has location information, and so on. As this process continues, the message migrates to the "best" node ("best" = least ID greater than or equal to B) in squares of increasing levels in the quad-tree at each step.

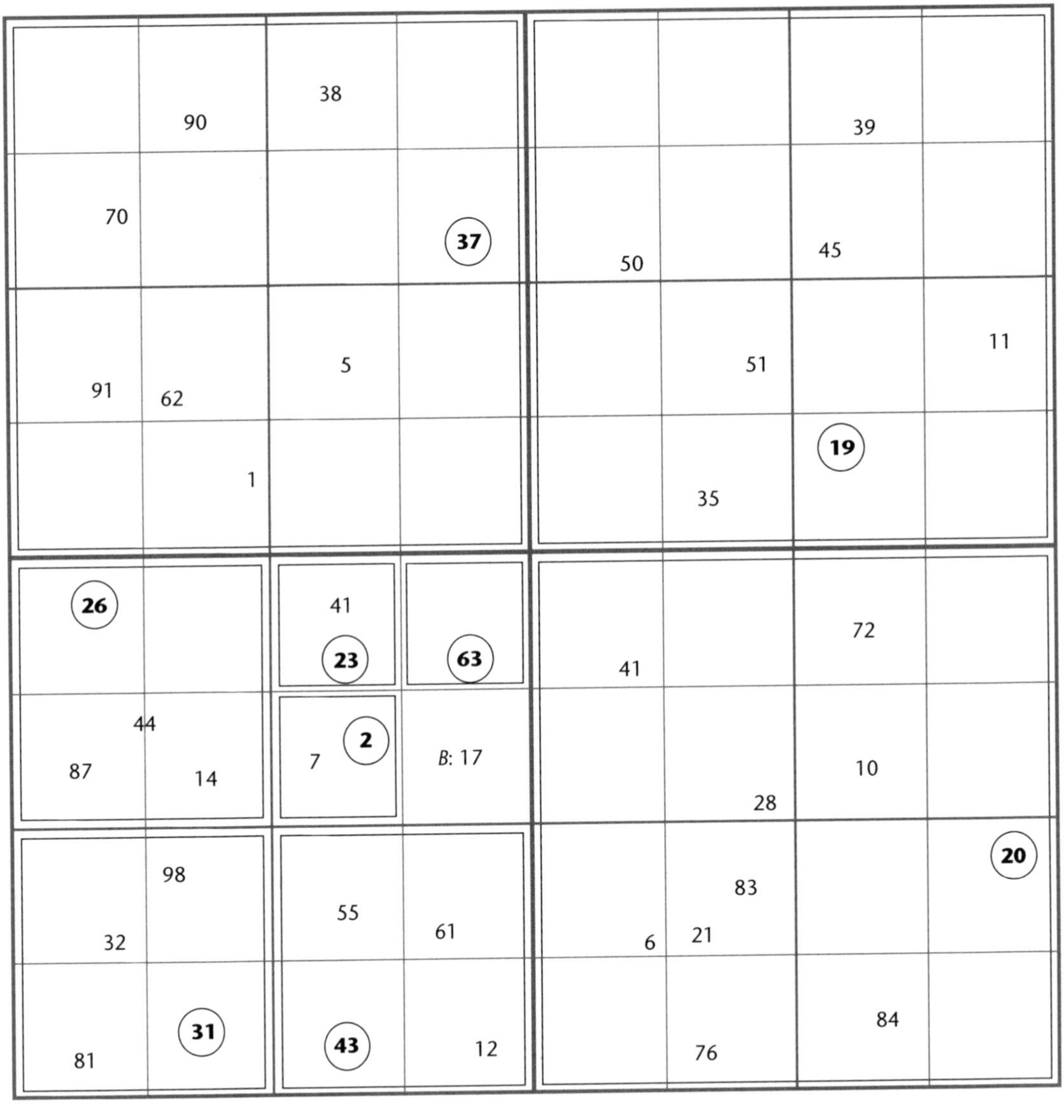

Figure 4.7 The location servers for node B, as selected by GLS (adapted from [134]).

At some point, the ascending tree path from A will meet the ascending path from B that established the triplets of B's location servers at each level. At that time, the search terminates and the location of B can be determined.

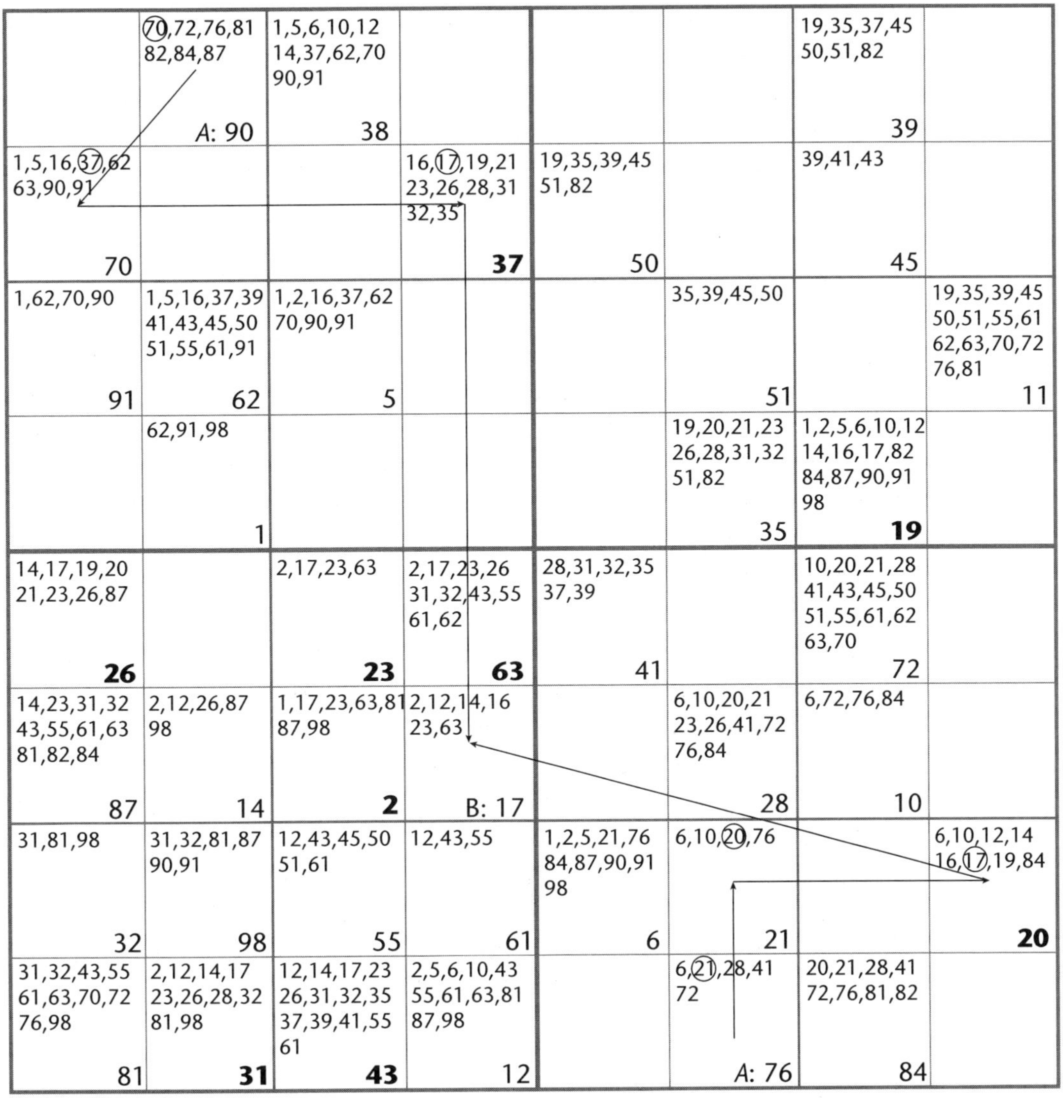

	70,72,76,81 82,84,87 A: 90	1,5,6,10,12 14,37,62,70 90,91 38				19,35,37,45 50,51,82 39	
1,5,16,37,62 63,90,91 70			16,17,19,21 23,26,28,31 32,35 37	19,35,39,45 51,82 50		39,41,43 45	
1,62,70,90 91	1,5,16,37,39 41,43,45,50 51,55,61,91 62	1,2,16,37,62 70,90,91 5			35,39,45,50 51		19,35,39,45 50,51,55,61 62,63,70,72 76,81 11
	62,91,98 1			19,20,21,23 26,28,31,32 51,82 35	1,2,5,6,10,12 14,16,17,82 84,87,90,91 98 19		
14,17,19,20 21,23,26,87 26		2,17,23,63 23	2,17,23,26 31,32,43,55 61,62 63	28,31,32,35 37,39 41		10,20,21,28 41,43,45,50 51,55,61,62 63,70 72	
14,23,31,32 43,55,61,63 81,82,84 87	2,12,26,87 98 14	1,17,23,63,81 87,98 2	2,12,14,16 23,63 B: 17		6,10,20,21 23,26,41,72 76,84 28	6,72,76,84 10	
31,81,98 32	31,32,81,87 90,91 98	12,43,45,50 51,61 55	12,43,55 61	1,2,5,21,76 84,87,90,91 98 6	6,10,20,76 21		6,10,12,14 16,17,19,84 20
31,32,43,55 61,63,70,72 76,98 81	2,12,14,17 23,26,28,32 81,98 31	12,14,17,23 26,31,32,35 37,39,41,55 61 43	2,5,6,10,43 55,61,63,81 87,98 12		6,21,28,41 72 A: 76	20,21,28,41 72,76,81,82 84	

Figure 4.8 The nodes for which a given node acts as a location server are shown in small type size in the quad-tree cell of that node. Those having B's location are shown in bold. The paths of two possible location queries for B are also shown, originating at node A (adapted from [134]).

The crucial observation here is that at each step of the search, the current node, say C, is associated with a tile at a certain level i of the quad-tree and forwards the message to a node D which is the best node in the parent tile of the tile containing C, at level $i + 1$. This is because node D must be known to C. Indeed, node D cannot lie in the same level i tile as D, because D is "better" than C. But D recruited location servers for itself at level i, including in the level i tile containing C. Since C is the smallest ID node greater than B in the tile, it is *a fortiori* the smallest ID node greater than D as well. Therefore, C was recruited by D, and D is known to C. Consider again the example in Figure 4.8. When node A is the one with ID 76, the query proceeds to nodes 21 and then 20, a node which is a location server for B. When node A is the one with ID 90, the query proceeds to nodes 70 and 37, where it terminates because again the latter is a location server for B.

This algorithm has many nice properties. In particular, if the source and destination lie in a common quad-tree tile at level i, then at most i steps are needed before a location server for the destination can be found for the source. This makes the cost of the location service distance-sensitive: the look-up time is sensitive to the separation between the source and destination. This is a reflection of the fact that a node selects more location servers near itself, and fewer and fewer as we move farther away. Thus in an area of the network where location servers for a node B are far from each other and where we may have to take many steps to find one, that is OK because node B itself is far away and the cost of reaching the server can be amortized over the cost of reaching the destination B.

4.5 Summary

In this chapter, we have touched upon many problems fundamental to setting up a sensor network when it is first deployed. The network topology must be established and radio ranges controlled (if that is possible) to balance adequate connectivity with efficient energy use. Nodes may be aggregated into clusters to better share resources, control redundancy, and enable hierarchical tasking and control. Furthermore, nodes need to have their clocks synchronized and locations determined, so that their subsequent detections and measurements can be placed into a common temporal and spatial framework for reasoning about the phenomena being observed.

A challenge common to the above problems is that they address global concerns across the entire network but they must be solved in a local, distributed manner. Although distributed protocols now exist for all these problems (and some of these have been presented here), it is fair to say that research in this area is still in its infancy. Since change is the norm in sensor networks, these protocols may have to be run not just once but repeatedly. Then energy use and load balancing become significant considerations that have not yet been well investigated. Synchronization and localization remain especially challenging problems in the ad hoc deployment setting we have focused on. Perhaps in some applications of sensor networks, additional infrastructure in the environment (e.g., a cellular network) can be assumed to be present to facilitate these tasks. This has been done in most earlier location-dependent systems, such as the *Cricket* indoor system developed at MIT [186], where beacons in each room broadcast location information to listener nodes.

Synchronization and localization can only be imperfectly implemented in most scenarios, yet they are central to many higher-level network tasks. Additional research to quantify the effects of temporal and spatial errors in performing these high-level operations is also clearly needed.

5

Sensor Tasking and Control

To efficiently and optimally utilize scarce resources in a sensor network, such as limited on-board battery power supply and limited communication bandwidth, nodes in a sensor network must be carefully tasked and controlled to carry out the required set of tasks while consuming only a modest amount of resources. For example, a camera sensor may be tasked to look for animals of a particular size and color, or an acoustic sensor may be tasked to detect the presence of a particular type of vehicle. To detect and track a moving vehicle, a pan-and-tilt camera may be tasked to anticipate and follow the vehicle object. It should be noted that to achieve scalability and autonomy, sensor tasking and control have to be carried out in a distributed fashion, largely using only local information available to each sensor.

For a given sensing task, as more nodes participate in the sensing of a physical phenomenon of interest and more data is collected, the total utility of the data, perhaps measured as the information content in the data, generally increases. However, doing so with all the nodes turned on may consume precious battery power that cannot be easily replenished and may reduce the effective communication bandwidth due to congestion in the wireless medium as well. Furthermore, as more nodes are added, the benefit often becomes less and less significant, as the so-called diminishing marginal returns set in, as shown in Figure 5.1. To address the balance between utility and resource costs, this chapter introduces a utility-cost-based approach to distributed sensor network management.

After discussing the general issues of task-driven sensing (Section 5.1), we develop a generic model of utility and cost (Section 5.2). Next, we present the main ideas of information-based sensor tasking

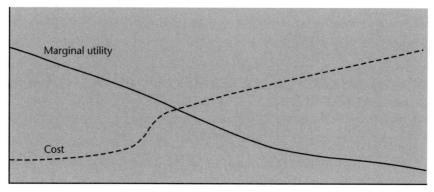

Number of nodes participating

Figure 5.1   Utility and cost trade-off: As the number of participating nodes increases, we see diminishing returns.

and discuss a specific realization in information-driven sensor querying (IDSQ) and a cluster-leader based protocol in which information about a physical phenomenon resides at a fixed leader for an extended period of time (Section 5.3). We then introduce dynamic migration of information within a sensor network, as in the case of tracking a moving phenomenon. Here, we will address the issues of information hand-off, greedy versus multistep lookahead information routing, as well as maintenance of collaborative processing sensor groups (Section 5.4). Although the material in this chapter is introduced in the context of localization or tracking tasks, the basic idea of information-based sensor tasking is applicable to more general sensing tasks.

5.1   Task-Driven Sensing

The purpose of a sensor system is often viewed as obtaining information that is as extensive and detailed as possible about the unknown parts of the world state. Any targets present in the sensor field need to be identified, localized, and tracked. All this data is to be centrally aggregated and analyzed. This is a reasonable view when the potential use of this information is not known in advance, and when the cost of the resources needed to acquire and transmit the information is either fixed or of no concern. Such a scheme, however, runs the danger of flooding the network with useless data and depleting scarce resources from battery power to human attention, as already mentioned. There are obvious ways to be more selective in choosing what sensor nodes to activate and what information to communicate; protocols such as directed diffusion routing (see Section 3.5.1) address exactly this issue for the transport layer of the network.

When we know the relevant manifest variables defining the world state—say, the position and identity of each target—then computing the answers to queries about the world state is a standard algorithm design problem. An algorithm typically proceeds by doing both numerical and relational (e.g., test) manipulations on these data, in order to compute the desired answer. The quality of the algorithm is judged by certain performance measures on resources, such as time and space used.

However, this classical algorithm/complexity view needs to be modified in the sensor network context because

- The values of the relevant manifest variables are not known, but have to be sensed.

- The cost of sensing different variables or relations of the same type can be vastly different—depending on the relative locations of targets and sensors, the sensing modalities available, the environmental conditions, and the communication costs.

- Frequently the value of a variable, or a relationship between variables, may be impossible to determine using the resources available in the sensor network; however, alternate variable values or relations may serve our purposes equally well.

Thus we need a new mathematical theory of algorithm design that includes the cost of accessing the manifest variables of the problem, or of determining useful atomic relationships among them. Furthermore, these costs cannot be precisely known in advance and may only be estimated. In addition, there may be values and relations that have been independently determined by the sensor network (say, while processing other tasks, or during a preprocessing step) which can be made available to the algorithm at a relatively low cost of communication. Thus the model needs to include both "push" and "pull" types of information flow.

To design an overall strategy, several key questions need to be addressed.

- What are the important objects in the environment to be sensed?

- What parameters of these objects are most relevant?

- What relations among these objects are critical to whatever high-level information we need to know?

- Which is the best sensor to acquire a particular parameter?

- How many sensing and communication operations will be needed to accomplish the task?

- How coordinated do the world models of the different sensors need to be?

- At what level do we communicate information, in the spectrum from signal to symbol?

Addressing these questions presents several challenges. Indeed, while the computational and communication complexity of different algorithms for the same problem can be assessed with standard techniques, the online nature of sensing requires the use of methods such as *competitive analysis* [212], *the value of information* [99], or other sensor utility measures [209, 28] to account for the fact that the value of sensor readings cannot be known before they are made.

5.2 Roles of Sensor Nodes and Utilities

Sensors in a network may take on different roles. Consider the following example of monitoring toxicity levels in an area around a chemical plant that generates hazardous waste during processing. A number of wireless sensors are initially deployed in the region [see Figure 5.2(a)]. Due to the nature of the environment and the cost of deployment, further human intervention or node replacement is not feasible. The sensors form a mesh network, and data collected by a subset of nodes is transmitted, through the multihop network,

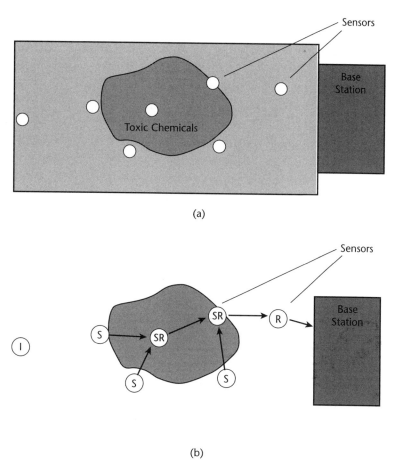

(a)

(b)

Figure 5.2 Sensor tasking: (a) A chemical toxicity monitoring scenario. (b) Sensors may take on different roles such as sensing (S), routing (R), sensing and routing (SR), or being idle (I), depending on the tasks and resources.

and relayed off the network to an adjoining base station or gateway. The network may be tasked to monitor the maximum toxicity levels in the region. To reduce the data traffic, individual toxicity detections from the sensors may be aggregated at an intermediate node, before being transmitted to the next node. In many cases, including this one, the aggregated data is of the same size as an individual detection.

A sensor may take on a particular role depending on the application task requirement and resource availability such as node power levels [Figure 5.2(b)]. Some of the nodes, denoted by SR in the figure, may participate in both sensing and routing. Note that routing includes both receiving and transmitting data. Some (S) may perform sensing only and transmit their data to other nodes. Some (R) may decide to act only as routing nodes, especially if their energy reserve is limited. Yet still others (I) may be in an idle or sleep mode, to preserve energy. As one can see, as the node energy reserve or other conditions change, a sensor may take on a different role. For example, a sensing-and-routing sensor may decide to drop the sensing role as soon as its energy reserve is below a certain level.

To study the problem of determining what role a sensor should play, we first introduce utility and cost models of sensors and then techniques that find optimal or nearly optimal assignments. A utility function assigns a scalar value, or utility, to each data reading of a sensing node; that is,

$$U : \mathcal{I} \times \mathcal{T} \to \mathcal{R}$$

where $\mathcal{I} = \{1, \ldots, K\}$ are sensor indices and $\mathcal{T}$ is the time domain. Each sensor operation is also assigned a cost. The cost of a sensing operation is C_s, aggregation cost is C_a, transmission cost is C_t, and reception cost is C_r. Note that these are unit costs per datum or packet, assuming the data in each operation can be so encapsulated. We further denote the set of nodes performing a sensing operation at time t as $V_s(t)$, aggregation nodes as $V_a(t)$, transmitting nodes as $V_t(t)$, and receiving nodes as $V_r(t)$. Omitting the issue of communication channel access contention and the possibility of retransmission (discussed in Section 3.2), we simplify the analysis by assuming the number of transmissions is the same as that of receptions within any period of time. We also omit the cost of listening.

As an instance of the constrained optimization problem introduced at the beginning of Chapter 2, the sensor tasking problem may be stated as follows:

Determine the sets of sensors V_s, V_t, V_r, and V_a that maximize the utility over a period of time

$$\max \sum_t \sum_{i \in V_s(t)} U(i, t),$$

subject to the constraint

$$\sum_t \sum_{V_s(t)} C_s + \sum_t \sum_{V_t(t)} (C_t + C_r) + \sum_t \sum_{V_a(t)} C_a \leq C_{total}.$$

We make a number of observations about the structure of utility and cost models. The utility of the network depends on the underlying routing structure. In a routing-tree realization, the tree must span the nodes that appear in the utility function and the base station. In a more general case, there may be multiple base stations or end-consumers of the data. The aggregate utility is assumed to be a monotonic function of the nodes participating in the sensing. We may further assume that the outcome of the aggregation operation is independent of the order in which the individual sensor readings are combined, in order to simplify the design of sensor tasking algorithms. Likewise, when the utility and costs of a sensing action are independent of the order of aggregation, efficient shortest-path algorithms can be used to find optimal routing paths.

These observations assume that the routing structure is statically determined a priori. However, in a task-driven sensing application, the amount of data to collect and aggregate often depends on the task goal and what information one already has, which may be dynamic. For example, the cardinality of the set V_s may be a function of the accuracy requirement of a localization task and the current configuration of available sensors. The order independence assumption in sensing utility and costs may be invalid when sensors contain redundant information, which is often the case in the sequential Bayesian estimation setting we discussed in Chapter 2. Consequently, the routing structure needs to adapt to the current sensing needs, and the cost of routing depends on factors such as sensor layout and how the data is aggregated. In this case, to determine the optimal allocation of sensing, aggregation, and communication resources, one may have to resort to more expensive search or approximation algorithms for exploring this larger combinatorial space of possibilities.

5.3 Information-Based Sensor Tasking

The main idea of information-based sensor tasking is to base sensor selection decisions on information content as well as constraints on resource consumption, latency, and other costs. Using information utility measures, sensors in a network can exploit the information content of data already received to optimize the utility of future sensing and communication actions, thereby efficiently managing scarce communication and processing resources. For example, IDSQ [233, 43] formulates the sensor tasking problem as a general distributed constrained optimization that maximizes information gain of sensors while minimizing communication and resource usage. We describe the main elements of the information-based approaches here.

5.3.1 Sensor Selection

Recall from Chapter 2 that for a localization or tracking problem, a belief refers to the knowledge about the target state such as position and velocity. In the probabilistic framework, this belief is represented as a probability distribution over the state space. We consider two scenarios, localizing a stationary source and tracking a moving source, to illustrate the use of information-based sensor tasking.

In the first scenario, a leader node might act as a relay station to the user, in which case the belief resides at this node for an extended time interval, and all information has to travel to this leader. In the second scenario, the belief itself travels through the network, and nodes are dynamically assigned as leaders. In this section, we consider the fixed leader protocol for the localization of a stationary source and postpone the discussion of the moving leader protocols to Section 5.4.

Given the current belief state, we wish to incrementally update the belief by incorporating the measurements of other nearby sensors. However, not all available sensors in the network provide useful information that improves the estimate. Furthermore, some information may be redundant. The task is to select an optimal subset and an optimal order of incorporating these measurements into our belief update. Note that in order to avoid prohibitive communication costs, this selection must be done without explicit knowledge of measurements residing at other sensors. The decision must be made solely based upon known characteristics of other sensors, such as their position and sensing modality, and predictions of their contributions, given the current belief about the phenomenon being monitored.

Figure 5.3 illustrates the basic idea of sensor selection. The illustration is based on the assumption that estimation uncertainty can be effectively approximated by a Gaussian distribution, illustrated by uncertainty ellipsoids in the state space. In the figure, the ellipsoid at time t indicates the residual uncertainty in the current belief state. The ellipsoid at time $t + 1$ is the incrementally updated belief after incorporating an additional sensor, either a or b, at the next time step. Although in both cases, a and b, the area of high uncertainty is reduced by the same percentage, the residual uncertainty in case a maintains the largest principal axis of the distribution. If we were to decide between the two sensors, we might favor sensor b over sensor a, based on the underlying measurement task.

Although details of the implementation depend on the network architecture, the fundamental principles derived in this chapter hold for both the selection of a remote sensor by a cluster-head as well

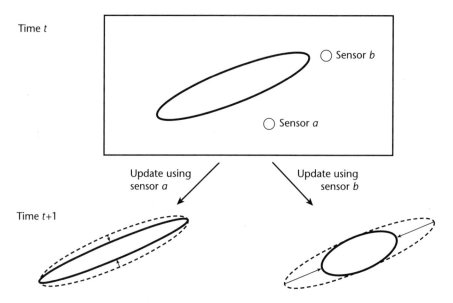

Figure 5.3   Sensor selection based on information gain of individual sensor contributions. The information gain is measured by the reduction in the error ellipsoid. In the figure, reduction along the longest axis of the error ellipsoid produces a larger improvement in reducing uncertainty. Sensor placement geometry and sensing modality can be used to compare the possible information gain from each possible sensor selection, *a* or *b*.

as the decision of an individual sensor to contribute its data and to respond to a query traveling through the network. The task is to select the sensor that provides the best information among all available sensors whose readings have not yet been incorporated. As will be shown in the experimental results, this provides a faster reduction in estimation uncertainty and usually incurs lower communication overhead for meeting a given estimation accuracy requirement, compared with blind or nearest-neighbor sensor selection schemes.

Example: Localizing a Stationary Source

In this example, 14 sensors are placed in a square region, as shown in Figure 5.4. Thirteen sensors are lined along the diagonal, with one sensor off the diagonal near the upper left corner of the square. The true location of the target is denoted by a cross in the figure.

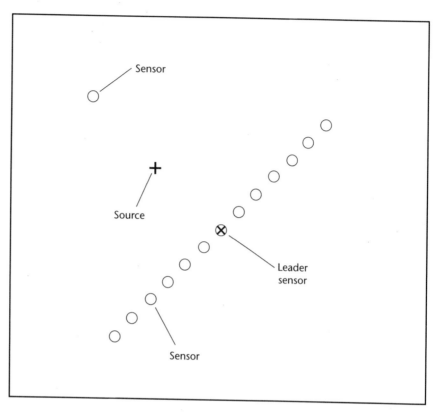

Figure 5.4 Source localization. The leader node (denoted by ×) queries other sensors (circles), to localize the source marked by a cross (+).

To illustrate sensor selection among one-hop neighbors, we assume all sensors can communicate to the node at the center, which queries all other nodes and acts as the data fusion center (the leader). We further assume that each sensor's measurement provides an estimate of the distance to the target, in the form of a doughnut-shaped likelihood function (see, e.g., Figure 2.7 of Chapter 2). The localization algorithm is based on sequential Bayesian estimation (2.8), which combines measurements from different sensors one at a time, assuming independence of likelihood functions when conditioned on the target state. Graphically, this amounts to taking the product of the doughnut-shaped likelihood functions from the sensors.

We first consider sensor selection based on a nearest-neighborhood (NN) criterion. Using this criterion, the leader node at the center always selects the nearest node among those whose measurements have not been incorporated. Figure 5.5 gives a sequence of snapshots of the localization results based on the NN criterion. Figure 5.5(a) shows the posterior distribution after incorporating the measurement from the initial leader sensor. Next, using the NN criterion, the best sensor is the next nearest neighbor in the linear array, and so forth [Figure 5.5(b)]. Figure 5.5(c) shows the resulting posterior distribution after the leader combines its data with the data from its two nearest neighbors in the linear array. The distribution remains as a bimodal distribution as data from additional sensors in the linear array are incorporated [Figure 5.5(d)-(e)], until the sensor at the upper-left corner of the sensor field is selected.

Alternately, we may select the next best sensor based on the so-called Mahalanobis measure, which captures the intuition developed in Figure 5.3 and whose definition will be given in the next section. Basically, the Mahalanobis-based selection favors sensors along the longer axis of the covariance fit of residual uncertainty in localization. For the first three measurements, the Mahalanobis-based method selects the same sequence of sensors as those by the NN-based method, as shown in Figure 5.5(a)-(c). After incorporating these measurements, the covariance fit of the residual uncertainty is elongated, and thus the upper-left sensor is selected as the next sensor according to the Mahalanobis measure. The new measurement from that sensor effectively reduces the current uncertainty to a more compact region [Figure 5.6(a); also compare Figure 5.6(a) with Figure 5.5(d)]. Next, the best sensor to select is the nearest one in the linear array [Figure 5.6(b)].

5.3.2 IDSQ: Information-Driven Sensor Querying

In distributed sensor network systems, we must balance the information contribution of individual sensors against the cost of communicating with them. For example, consider the task of selecting among K sensors with measurements $\{z_i\}_{i=1}^{K}$. Given the current belief $p(\mathbf{x} \mid \{z_i\}_{i \in U})$, where $U \subset \{1, \ldots, K\}$ is the subset of sensors whose measurement has already been incorporated, the task is to

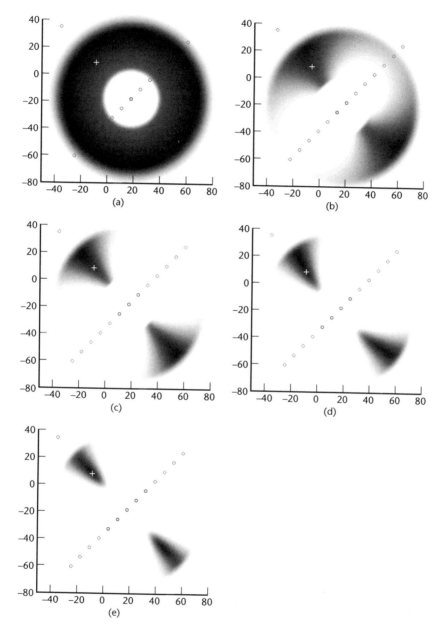

Figure 5.5 Sensor selection based on the nearest-neighbor method. The estimation task here is to localize a stationary target labeled "+". Circles denote sensors, and thick circles indicate those whose measurements have already been incorporated. (a) Residual uncertainty after incorporating the data from the leader at the center. (b)-(e) Residual uncertainty after incorporating each additional measurement from a selected sensor.

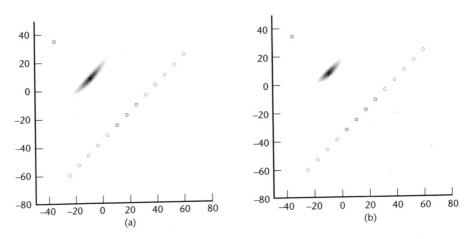

Figure 5.6 Sensor selection based on the Mahalanobis measure of information utility. The localization problem is the same as that in Figure 5.5. The residual uncertainties shown represent the results after incorporating measurements from the 4th and 5th sensors, respectively.

determine which sensor to query among the remaining unincorporated set $A = \{1, \ldots, K\} - U$. This is a reasonable strategy for localizing a stationary target. For moving targets, the same sensor may provide informative measurements at different times. The problem of tracking moving targets is discussed in Section 5.4.3.

To be precise, let us define an information utility function

$$\phi : \mathcal{P}(\mathcal{R}^d) \to \mathcal{R}.$$

$\mathcal{P}(\mathcal{R}^d)$ represents the class of all probability distributions on d-dimensional state space $\mathcal{R}^d$ for the target state $\mathbf{x}$. The utility function ϕ assigns a scalar value to each element $p \in \mathcal{P}(\mathcal{R}^d)$, which indicates how spread out or uncertain the distribution p is. Smaller values represent a more spread out distribution, while larger values represent a tighter, lower-variance distribution. Different choices of ϕ will be discussed later in the section. We further define a cost of obtaining a measurement as a function:

$$\psi : \mathcal{R}^h \to \mathcal{R}$$

where $\mathcal{R}^h$ is an h-dimensional measurement space where a measurement vector lies.

In the following, we will refer to sensor l, which holds the current belief, as the *leader node*. The constrained optimization problem of sensor tasking can be reformulated as an unconstrained optimization problem, with the following objective function as a mixture of information utility and cost:

$$J\left(p(\mathbf{x}\,|\,\{\mathbf{z}_i\}_{i\in U} \cup \{\mathbf{z}_j\})\right)$$
$$= \gamma \cdot \phi\left(p\left(\mathbf{x}\,|\,\{\mathbf{z}_i\}_{i\in U} \cup \{\mathbf{z}_j\}\right)\right) - (1-\gamma) \cdot \psi\left(\mathbf{z}_j\right). \qquad (5.1)$$

Here ϕ measures the information utility of incorporating the measurement $\mathbf{z}_j^{(t)}$ from sensor $j \in A$, ψ is the cost of communication and other resources, and γ is the relative weighting of the utility and cost. It should be noted that ϕ could measure either the total information utility of the belief state after incorporating the new measurement or just the incremental information gain, whichever is easier to compute. With this objective function, the sensor selection criterion takes the form

$$\hat{j} = \arg\max_{j\in A} J\left(p\left(\mathbf{x}\,|\,\{\mathbf{z}_i\}_{i\in U} \cup \{\mathbf{z}_j\}\right)\right). \qquad (5.2)$$

However, in practice, we do not know the measurement value $\mathbf{z}_j$ without transmitting it to the current aggregation center, the node l, first. Nevertheless, we wish to select the "most likely" best sensor, based on the current belief state $p(\mathbf{x}\,|\,\{\mathbf{z}_i\}_{i\in U})$ plus our knowledge of the measurement model and sensor characteristics. For example, the cost function ψ may be estimated as the distance between sensor j and sensor l, or the distance raised to some power, as a rough indicator of how expensive it is to transmit the measurement. As the result, we often compute an estimate of the cost, $\hat{\psi}$, from parameters such as λ_j and λ_l, the sensor characteristics introduced in Section 2.2.1. On the other hand, the utility function ϕ cannot be computed without the measurement $\mathbf{z}_j$. Instead, we compute an estimate of the utility, $\hat{\phi}$, by marginalizing out the particular value of $\mathbf{z}_j$. Note that for any given value of $\mathbf{z}_j$ for sensor j, we get a particular value for ϕ acting on the new belief state $p(\mathbf{x}\,|\,\{\mathbf{z}_i\}_{i\in U} \cup \{\mathbf{z}_j\})$. Now, for each sensor j, consider the set of all values of ϕ for different choices of $\mathbf{z}_j$. Possibilities for summarizing the set of values of ϕ by a single quantity include considering the average, the worst, or the best case. Detailed discussions of these can be found in [43].

In the following, we use the approximations $\hat{\phi}$ and $\hat{\psi}$ whenever we discuss utility and cost. Further abusing the notation, the arguments to $\hat{\phi}$ and $\hat{\psi}$ are not fixed, and the approximation functions may take various forms, depending on the application and context.

Information Utility Measure I: Mahalanobis Distance

Assume the belief state is well approximated by a Gaussian distribution, and sensor data provide a range estimate, as in acoustic amplitude sensing, for example. In the sensor configuration shown in Figure 5.3, sensor a would provide better information than b because sensor a lies close to the longer axis of the uncertainty ellipsoid and its range constraint would intersect this longer axis transversely. To favor the sensors along the longer axes of an uncertainty ellipsoid, we use the Mahalanobis distance, a distance measure normalized by the uncertainty covariance Σ. The (squared) Mahalanobis distance from $\mathbf{y}$ to μ is defined as

$$(\mathbf{y} - \mu)^T \Sigma^{-1} (\mathbf{y} - \mu).$$

The utility function for a sensor j, with respect to the target position estimate characterized by the mean $\hat{\mathbf{x}}$ and covariance Σ, is defined as the negative of the Mahalanobis distance

$$\hat{\phi}\left(\zeta_j,\ \hat{\mathbf{x}},\ \Sigma\right) = -\left(\zeta_j - \hat{\mathbf{x}}\right)^T \Sigma^{-1} \left(\zeta_j - \hat{\mathbf{x}}\right), \tag{5.3}$$

where ζ_j is the position of sensor j.

Intuitively, the points on the 1-σ surface of the error covariance ellipsoid are all equidistant from the center under the Mahalanobis measure (Figure 5.7). The utility function works well when the current belief can be well approximated by a Gaussian distribution or

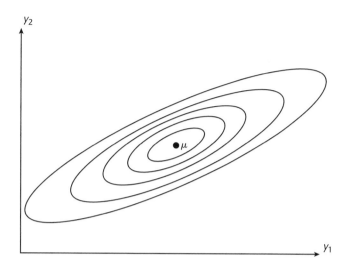

Figure 5.7 Mahalanobis measure: Points on a constant density contour of a Gaussian distribution $N(\mu, \Sigma)$ are equidistant from the mean μ.

the distribution is very elongated, and the sensors are range sensors. However, a bearing sensor reduces the uncertainty along the direction perpendicular to the target bearing and requires a different measure of utility. For a general uncertainty distribution or bearing sensors, we must develop alternative information utility measures.

Information Utility Measure II: Mutual Information

For multimodal, non-Gaussian distributions, a mutual information measure provides a better characterization of the usefulness of sensor data. Additionally, this measure is not limited to information based only on range data. Assume the current belief is $p(\mathbf{x}^{(t)}|\overline{\mathbf{z}^{(t)}})$. The contribution of a potential sensor j is measured by

$$\hat{\phi}\left(\mathbf{z}^{(t)},\, p\left(\mathbf{x}^{(t)}\big|\overline{\mathbf{z}^{(t)}}\right)\right) = I\left(X^{(t+1)};\, Z_j^{(t+1)}\big|\overline{Z^{(t)}} = \overline{\mathbf{z}^{(t)}}\right), \qquad (5.4)$$

where $I(\cdot\,;\cdot)$ measures the mutual information in bits between two random variables. Essentially, maximizing the mutual information is equivalent to selecting a sensor whose measurement $\mathbf{z}_j^{(t+1)}$, when conditioned on the current measurement history $\overline{\mathbf{z}^{(t)}}$, would provide the greatest amount of new information about the target location $\mathbf{x}^{(t+1)}$. The mutual information can be interpreted as the Kullback-Leibler divergence between the belief after and before applying the new measurement $\mathbf{z}_j^{(t+1)}$. Therefore, this criterion favors the sensor that, on average, gives the greatest change to the current belief. An implementation of a real-time tracking system using this utility function has shown that this measure is both practically useful and computationally feasible [144].

Appendix C at the end of the book develops additional forms of utility measures. The appropriateness of a particular utility measure for a sensor selection problem depends on two factors: the characteristics of the problem, such as the data and noise models, and the computational complexity of computing the measure. For example, the Mahalanobis measure is easy to compute, although limited to certain data models. The mutual information applies to multimodal distributions, but its computation requires expensive convolution of discrete points if one uses a grid approximation of probability density functions. The choice of which measures to use illustrates the important design trade-off for sensor networks: optimality in information versus feasibility in practical implementation.

5.3.3 Cluster-Leader Based Protocol

The IDSQ method is based on the cluster-leader type of distributed processing protocol. Although the algorithm presented here assumes there is a single belief carrier node active at a time, the basic ideas also apply to scenarios where multiple belief carriers are active simultaneously, as long as the clusters represented by the belief carriers are disjointed from each other; in other words, each sensor senses a single target at a time. Assume we have a cluster of K_1 sensors, each labeled by the integer $\{1, \ldots, K_1\}$. A priori, each sensor i only has knowledge of its own position $\zeta_i \in \mathcal{R}^2$. An important prerequisite is the appropriate cluster formation and leader election before applying the algorithm. The cluster may be initially formed from sensors with detections above a threshold and updated as the signal source moves. A leader may be elected based on relative magnitude of measurement or time of detection (discussed later in this section). Techniques for clustering and leader election are also discussed in Section 4.2 of Chapter 4.

We develop an IDSQ algorithm, using an information criterion for sensor selection and Bayesian filtering for data fusion (see Section 2.2.3), in the context of localization tasks. As pointed out earlier, the basic algorithm introduced here should be equally applicable to other sensing problems. Figure 5.8 shows the flowchart of this algorithm which is identical for every sensor in the cluster. The algorithm works as follows:

Initialization (Step 1): Each sensor runs an initialization routine through which a leader node is elected from a cluster of K_1 sensors who have detections. The leader election protocol will be considered later. Other sensors in the cluster communicate their own characteristics $\{\lambda_i\}_{i=1}^{K_1}, (i \neq l)$, as defined in Section 2.2.1, which include the position and noise variance of each sensor, to the leader l.

Follower Nodes (Step 2a): If the sensor node is not the leader, then the algorithm follows the left branch in Figure 5.8. These nodes will wait for the leader node to query them, and if they are queried, they will process their measurements and transmit the queried information back to the leader.

Initial Sensor Reading (Step 2b): If the sensor node is the leader, then the algorithm follows the right branch in Figure 5.8. The leader node will then

1. calculate a representation of the belief state with its own measurement, $p(\mathbf{x} \mid \mathbf{z}_l)$, and

2. begin to keep track of which sensor measurements have been incorporated into the belief state, $U = \{l\} \subset \{1, \dots, K_1\}$.

Again, it is assumed that the leader node has knowledge of the characteristics $\{\lambda_i\}_{i=1}^{K_1}$ of all the sensors within the cluster.

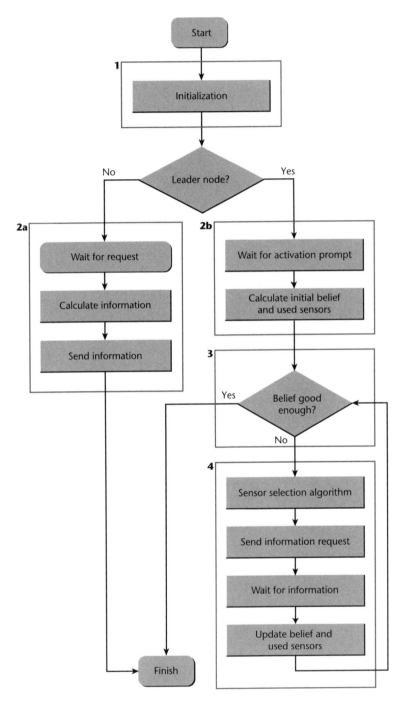

Figure 5.8 Flowchart of the information-driven sensor querying (IDSQ) algorithm for each sensor (adapted from [43]).

Belief Quality Test (Step 3): If the belief is good enough, based on some measure of goodness such as the size of belief, the leader node is finished processing. Otherwise, it will continue with sensor selection.

Sensor Selection (Step 4): Based on the belief state, $p(\mathbf{x} \mid \{\mathbf{z}_i\}_{i \in U})$, and sensor characteristics, $\{\lambda_i\}_{i=1}^{K_1}$, pick a sensor node from $\{1, \ldots, K_1\} - U$ that satisfies some information criterion ϕ, assuming the one-hop cost is identical for all sensors. Say that node is j. Then, the leader will send a request for sensor j's measurement, and when the leader receives the requested information, it will

1. update the belief state with $\mathbf{z}_j$ to get a representation of

$$p(\mathbf{x} \mid \{\mathbf{z}_i\}_{i \in U} \cup \mathbf{z}_j), \text{ and}$$

2. add j to the set of sensors whose measurements have already been incorporated:

$$U := U \cup \{j\}.$$

Now, loop back to step 3 until the belief state is good enough.

At the end of this algorithm, the leader node contains all the information about the belief from the sensor nodes by intelligently querying a subset of the nodes that provide the majority of the information. This reduces unnecessary power consumption by transmitting only the most useful information to the leader node. This computation can be thought of as a local computation for this cluster. The belief stored by the leader can then be passed up for processing at higher levels. In steps 2b and 4, some form of representation of the belief $p(\mathbf{x} \mid \{\mathbf{z}_i\}_{i \in U})$ is stored at the leader node. Considerations for the particular representation of the belief was mentioned in Section 2.3. In step 4, an information criterion is used to select the next sensor. Different measures of information utility, were discussed in Section 5.3.2 and then in details in Appendix C at the end of the book.

Leader Election Protocol

In the leader-based protocol, it is necessary to design efficient and robust algorithms for electing a leader, since typically more than one sensor may have detections about a target simultaneously. Here, we describe a geographically-based leader election scheme that resolves contention and elects a single leader via message exchange.

First, consider the ideal situation. If the signal of a target propagates isotropically and attenuates monotonically with distance, the sensors physically closer to the target are more likely to detect the target than the sensors far away. One can compute an "alarm region," similar to a 3-σ region of a Gaussian distribution, such that most (e.g., 99 percent) of the sensors with detections fall in the region. This is illustrated in Figure 5.9. Sensor nodes are marked with small circles; the dark ones have detected a target. Assume the target is located at **x** (marked with a "+" in the figure), the alarm region is a

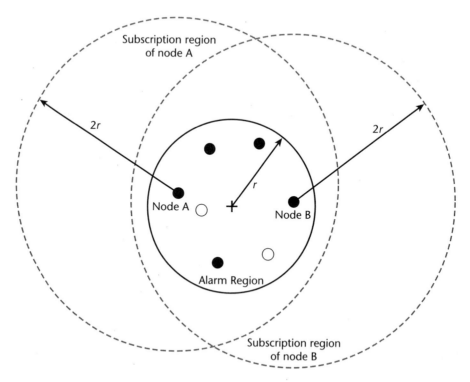

Figure 5.9 Leader election in a detection region (adapted from [142]).

disk centered at $\mathbf{x}$ with some radius r, where r is determined by the observation model. In practice, we choose r based on the maximum detection range plus a moderate amount of margin to account for possible target motion during a sample period.

Ideally, nodes in an alarm region should collaborate to resolve their contention and elect a single leader from that region. However, the exact location of the alarm region is unknown since the target position $\mathbf{x}$ is unknown. Each node with a detection only knows that the target is within a distance of r, and a possible competitor could be a farther distance r from the target. Thus in the absence of a "message center," a node notifies all nodes within a radius $2r$ of itself (the potential "competitors" for leadership) of its detection.

Upon detection, each node broadcasts to all nodes in the enlarged alarm region a DETECTION message containing a time stamp indicating when the detection is declared, and the likelihood ratio $p(\mathbf{z}|H_1)/p(\mathbf{z}|H_0)$, where H_1 or H_0 denote the hypotheses of the target being present or not. The higher this ratio, the more confident the detecting node is of its detection. We rely on a clock synchronization algorithm to make all time stamps comparable, as discussed in Section 4.3. We also need a routing mechanism to effectively limit the propagation of the detection messages to the specified region, using, for example, the geographical routing to a region method presented in Section 3.4.4.

After sending out its own detection message, the node checks all detection packets received within an interval of t_{comm}. The value of t_{comm} should be long enough for all messages to reach their destination, yet not too long so that the target can be considered approximately stationary. These messages are then compared with the node's own detection. The node winning this election then becomes the leader immediately, with no need for further confirmation. The election procedure is as follows:

- If none of the messages is time-stamped earlier than the node's own detection, the node declares itself leader.

- If there are one or more messages with an earlier time stamp, the node knows that it is not the leader.

• If none of the messages contains earlier time stamps, but some message contains a time stamp identical to the node's detection time, the node compares the likelihood ratio. If the node's likelihood ratio is higher, the node becomes the leader.

This algorithm elects only one leader per target in an ideal situation. However, in other situations multiple leaders may result. For example, if the DETECTION packet with the earliest detection time stamp fails to reach all the destination nodes, multiple nodes may find that they are the "earliest" detection and each may initiate a localization task. One way to consolidate the multiple leaders is to follow the initial election with another round of election, this time involving only the elected leaders from the initial round. Since there are fewer nodes to send out messages, the probability of DETECTION packets reaching every leader is greatly increased.

Simulation Experiments

The leader-based protocol is applied to the problem of spatial localization of a stationary target based on amplitude measurements from a network of 14 sensors, as arranged in Figure 5.4. The goal is to compare different sensor selection criteria and validate the IDSQ algorithm presented earlier. Assuming acoustic sensors, the measurement model for each sensor i is given in Equation (2.4) of Chapter 2. For the experiment, we further assume $a \in \mathcal{R}$, the source amplitude of the target signal, is uniformly distributed in the interval $[a_{low}, a_{high}]$. For simplicity, in the simulation examples considered in this section, a leader is chosen to be the one whose position ζ_l is closest to the centroid of the sensors, that is,

$$l = \arg \min_{j=1,\ldots,N} \left\| \zeta_j - \frac{1}{N} \sum_{i=1}^{N} \zeta_i \right\|.$$

We test four different criteria for choosing the next sensor:

A. Nearest neighbor

B. Mahalanobis distance

C. Maximum likelihood

D. Best feasible region (or ground truth for one-step optimization)

Criterion D is uncomputable in practice since it requires knowledge of sensor measurement values in order to determine which sensor to use. However, we use it as a basis for comparison with the other criteria. These four criteria are defined more precisely in Appendix D at the end of the book.

Additional details of the protocol and sensor selection criteria are specified as follows. For the simulations, we have chosen $a_{low} = 10$ and $a_{high} = 50$. The sensor noise variance σ_i is set to 0.1, which is about 10 percent of the signal amplitude when the amplitude of the target is 30 and the target is at a distance of 30 units from the sensor. The parameter β in criterion D (as specified in Appendix D) is chosen to be 2, since this value covers 99 percent of all possible noise instances. For the first simulation, the signal attenuation exponent α is set to 1.6, which considers reflections from the ground surface. Then α is set to 2 for the second simulation, which is the attenuation exponent in free space with no reflections or absorption. The shape of the uncertainty region is sensitive to different choices of α; however, the comparative performance of the sensor selection algorithm for different selection criteria turns out to be relatively independent of α.

The first simulation is a pedagogical example to illustrate the usefulness of incorporating a sensor selection algorithm into the sensor network. Figure 5.4 earlier in the chapter shows the layout of 14 microphones. The one microphone not in the linear array is placed so that it is farther from the leader node than the farthest microphone in the linear array. With sensor measurements generated by a stationary source in the middle of the sensor network, sensor selection criteria A and B are compared. The difference in the two criteria is the order in which the sensors' measurements are incorporated into the belief.

Figure 5.10 shows a plot of the logarithm of the determinant of the error covariance of the belief state (or the volume of the error

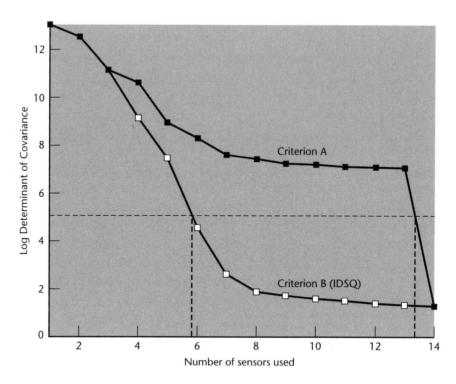

Figure 5.10 Determinant of the error covariance for selection criteria A and B for the sensor layout shown in Figure 5.4. Criterion A tasks 14 sensors, while B tasks 6 sensors to be below an error threshold of five units (adapted from [43]).

ellipsoid) versus the number of sensors incorporated. Indeed, the volume of the error covariance under selection criterion B is less than the volume of the error covariance under selection criterion A for the same number of sensors, after more than three sensors are selected. When all 14 sensors have been accounted for, both methods produce the same amount of error reduction.

A plot of the communication distance versus the number of sensors incorporated is shown in Figure 5.11. Certainly, the curve for selection criterion A is the lower bound for any other criterion. Criterion A optimizes the network to use the minimum amount of communication energy when incorporating sensor information; however, it largely ignores the information content of these sensors.

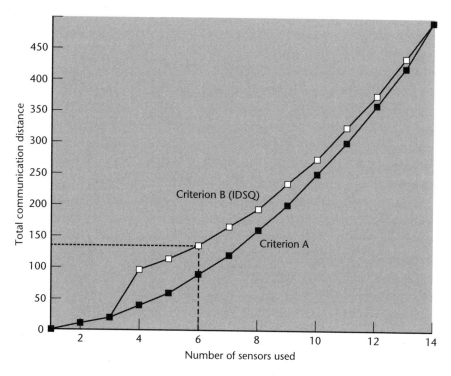

Figure 5.11 Total communication distance versus the number of sensors queried for selection criteria A and B for the sensor layout shown in Figure 5.4. For achieving the same threshold of the error, A tasks 14 sensors and uses nearly 500 units of communication distance, whereas B tasks 6 sensors and uses less than 150 units of communication distance (adapted from [43]).

A more informative interpretation of the figure is to compare the amount of energy it takes for criterion A and criterion B to achieve the same level of accuracy. Examining Figure 5.10, we see that under criterion A, in order for the log determinant of the covariance value to be less than 5, criterion A requires all 14 sensors to be tasked. On the other hand, criterion B requires only 6 sensors to be tasked. Now, comparing the total communication distance for this level of accuracy from Figure 5.11, we see that criterion B requires less than 150 units of communication distance for tasking 6 sensors, as opposed to nearly 500 units of communication distance for tasking all 14 sensors. Indeed, for a given level of accuracy, B generally requires less communication distance than A.

The above simulation was carried out on a specific layout of the sensors, and the striking improvement of the error was largely due to the fact that most of the sensors were in a linear array. Thus, the next simulation will explore which one does better, on average, with randomly placed sensors.

Microphones are placed uniformly in a given square region as shown in Figure 5.12(a). The target is placed in the middle of the square region and given a random amplitude. Then, the sensor algorithm for each of the different sensor selection criteria described earlier is run for 200 runs. Figure 5.12(b) shows a comparison between selection criteria A and B. There are three segments in each bar. The bottom segment represents the percentage of runs in which the error for B is strictly less than the error for A after k sensors have been incorporated. The middle represents a tie. The upper segment represents the percentage of runs in which the error for B is larger than the error for A. Since the bottom segment is larger than the upper one (except for the initial and final phases when they are tied), this shows B performs better than A on average.

Figures 5.12(c) and (d) show comparisons of sensor criteria C and D versus B. The performance of C is comparable to B and, as expected, D is better than B on average. The reason D is not always better than B over *a set of sensors* is because both are greedy criteria. The nth best sensor is chosen incrementally with the first $n-1$ sensors already fixed. Fixing the previous $n-1$ sensors when choosing the nth sensor is certainly suboptimal to choosing n sensors all at once to maximize the information content of the belief.

5.3.4   Sensor Tasking in Tracking Relations

The focus of this section is sensor tasking and control in the context of tracking spatial or temporal relations between objects and local or global attributes of the environment—as opposed to the detailed estimation of positions and poses of individual objects. In certain cases, high-level behaviors of objects may be trackable more robustly than their exact positions, relations between objects may be directly

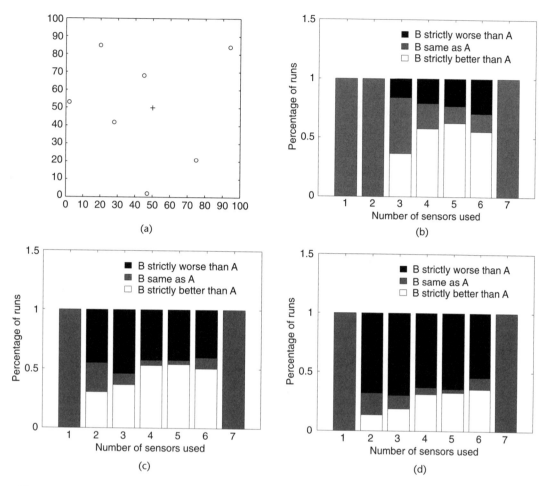

Figure 5.12 (a) Layout of seven randomly placed sensors (circles) with the target (cross) in the middle. (b) Percentage of runs where B performs better than A for seven randomly placed sensors. (c) Percentage of runs where B performs better than C for seven randomly placed sensors. (d) Percentage of runs where B performs better than D for seven randomly placed sensors (adapted from [43]).

observable by sensors, and the large-scale behavior of an ensemble of objects may be easier to ascertain than the motion of the individual objects. By focusing on relations and the logical structure of the evidence with respect to the task at hand, information-based approaches will be able to allocate sensing, computation, and communication resources where they are most needed.

An example of tracking relations is the "Am I surrounded?" problem: determine if a friendly vehicle is surrounded by a number of enemy tanks [Figure 5.13(a)]. The goal is to design a sensing strategy that extracts global relations among the vehicles in question without first having to solve local problems, such as accurately localizing individual vehicles. One definition of whether the friendly vehicle is surrounded by the tanks is to test if the vehicle is inside the geometric convex hull formed by the enemy tanks. Although the notion of "Am I surrounded?" is somewhat application dependent, we will use this definition to show how such a global relation can be determined by tasking appropriate sensors, based on how their local sensing actions can contribute to the assertion of the relation. We start by decomposing a global relation into more primitive ones. For example, the global relation of whether a point d is surrounded by points a, b, c can be established by the conjunction of the more primitive relations $CCW(a, b, d)$, $CCW(b, c, d)$, $CCW(c, a, d)$, where CCW denotes the counterclockwise relation [Figure 5.13(b)].

To establish a CCW relation among three objects, sensors are selected to localize the objects, with the objective of maximizing

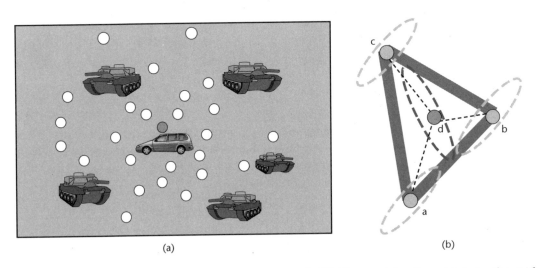

(a)   (b)

Figure 5.13 (a) A global relation of "Am I surrounded?" Here, a friendly vehicle in the middle is attempting to determine if it is inside the convex hull of enemy tanks. (b) Decomposition of a global relation into three more primitive CCW relations.

the reduction in the uncertainty of the CCW relation while minimizing the number of sensor queries. To resolve a CCW relation, a
sensor with maximum expected reduction in the uncertainty of the
CCW relation is chosen. For example, to resolve the CCW(a, b, d)
in Figure 5.13(b), where ellipses denote uncertainty covariances for
localization of a, b, and d, one notices that tasking sensors that minimize the error for node d can result in faster resolution of the CCW
relation. Figure 5.14 shows a simulation of the CCW resolution as

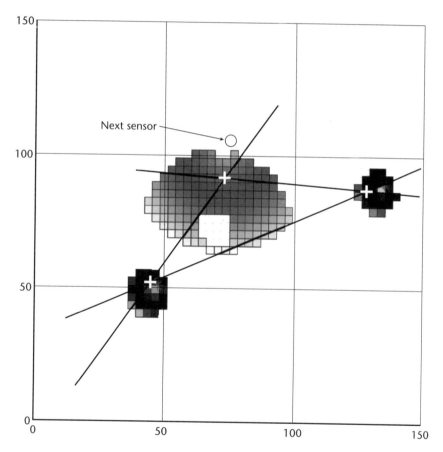

Figure 5.14 A CCW relation is being resolved by tasking sensors with maximum expected
information gain. A "+" denotes the true location for each target (adapted from [87]).

sensors are selected. In the figure, each node is localized with a set of range sensors, resulting in residual uncertainties about object positions shown in the figure.

Note that uncertainty in the CCW relation is caused by possible collinearity or near-collinearity of the targets. The best sensor selection strategy for removing collinearity uncertainty may be quite different from the best strategy for localizing the sensors, irrespective of the CCW relation. Consider, for instance, the situation in Figure 5.15. Note that a passive infrared (PIR) sensor s_1 may look at one of the tails of the distribution for target t_3 and, upon seeing nothing there, lop off a large chunk of this distribution and reduce its spread. Yet that reduction is almost useless as far as eliminating wrongly oriented triplets of possible target locations. Another PIR sensor s_2 may lop off a smaller part of the t_2 distribution, yet have a much more significant benefit toward certifying CCW(t_1, t_2, t_3). Analogous to the sensor selection for the point localization problems, we need to develop a model of utility and cost that relates the resolution of global relations to the sensing and communications required. This remains as an open problem for future research.

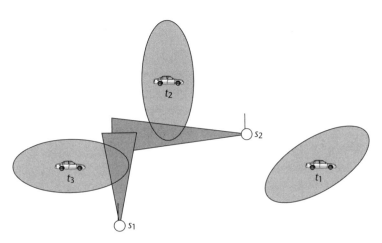

Figure 5.15 The effect on CCW of alternate sensor readings (adapted from [87]).

5.4 Joint Routing and Information Aggregation

A primary purpose of sensing in a sensor network is to collect and aggregate information about a phenomenon of interest. While IDSQ provides us with a method of selecting the optimal order of sensors to obtain maximum incremental information gain, it does not specifically define how a query is routed through the network or the information is gathered along the routing path. This section outlines a number of techniques that exploit the composite objective function (5.1) to dynamically determine the optimal routing path.

Consider the following two scenarios in which an information aggregation query is injected into the network. The first one is illustrated in Figure 5.16(a). A user (which may be another sensor node)

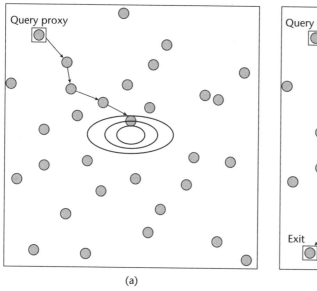
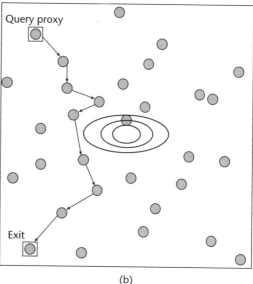

(a)   (b)

Figure 5.16 Routing and information aggregation scenarios: (a) Routing from a query proxy to the high activity region and back. The ellipses represent iso-contours of an information field, which is maximal at the center. The goal of routing is to maximally aggregate information along a path while keeping the cost minimal. (b) Routing from a query proxy to an exit node, maximizing information gain along the path (adapted from [145]).

issues a query from an arbitrary node, which we call a *query proxy node*, requesting the sensor network to collect information about a phenomenon of interest. The query proxy has to figure out where such information can be collected and routes the query toward the high information content region. This differs from routing in communication networks where the destination is often known a priori to the sender. Here, the destination is unknown and is dynamically estimated by the network's knowledge about the physical phenomenon.

The second routing scenario is pictured in Figure 5.16(b). A user—for example, an police officer—may issue a query to a node, asking the sensor network to collect information and report the result to an extraction or exit node—for example, a police station—where the information can be extracted for further processing. In this scenario, the query proxy and exit nodes may be far away from the high information content region. A path taking a detour toward the high information region may be more preferable than the shortest path. In this section, we will first consider the sensor tasking and routing problem for the first scenario. The problems associated with the second scenario are studied in Section 5.4.2.

5.4.1 Moving Center of Aggregation

We have described a class of algorithms based on the fixed belief carrier protocol in which a designated node such as a cluster leader holds the belief state. In that case, the querying node selects optimal sensors to request data from, using the information utility measures. For example, using the Mahalanobis distance measure, the querying node can determine which node can provide the most useful information while balancing the communication cost, without the need to obtain the remote sensor data first. We now consider a dynamic belief carrier protocol in which the belief is successively handed off to sensor nodes closest to locations where "useful" sensor data are being generated. In the dynamic case, the current sensor node updates the belief with its measurement and sends the estimation to the next neighbor that it determines can best improve the estimation.

Locally Optimal Search

Here the information query is directed by local decisions of individual sensor nodes and guided into regions maximizing the objective function J as defined in (5.1). Note that the function J is incrementally updated along with the belief updates along the routing path. The local decisions can be based on different criteria:

1. For each sensor k that makes the current routing decision, evaluate the objective function J at the positions of the m sensors within a local neighborhood determined by the communication distance, and pick the sensor j that maximizes the objective function locally within the neighborhood:

$$\hat{j} = \arg\max_j (J(\zeta_j)), \quad \forall j \neq k$$

where ζ_j is the position of the node j.

2. Choose the next routing sensor in the direction of the gradient of the objective function, ∇J. Among all sensors within the local communication neighborhood, choose the sensor j such that

$$\hat{j} = \arg\max_j \left(\frac{(\nabla J)^T \cdot (\zeta_j - \zeta_k)}{\|\nabla J\| \, \|\zeta_j - \zeta_k\|} \right),$$

where ζ_k is the position of the current routing node, and "$\cdot$" denotes the inner product of two vectors.

3. If the network routing layer supports geographical routing, as described in Section 3.4, the querying sensor can directly route the query to the sensor closest to the optimum position. The optimum position ζ_o corresponds to the location where the utility function ϕ is maximized and can be computed by the querying sensor by evaluating the utility function:

$$\zeta_o = \arg_\zeta [\nabla \phi = 0]. \tag{5.5}$$

However, the destination is optimal only with respect to the current data the querying sensor has. As the query travels through the sensor nodes in the network, additional sensor data may be incrementally combined to continuously update the optimum position.

4. Instead of following the local gradients of the objective function throughout the routing path, the chosen direction at any hop can be biased toward the direction aiming at the optimum position, ζ_o. This variation of the gradient ascent algorithm is most useful in regions of small gradients of the objective function, that is, where the objective function is relatively flat. The direction toward the maximum of the objective function can be found by evaluating (5.5) at any routing step. This allows a node to compute locally the direction toward the optimum position $(\zeta_o - \zeta_k)$, where ζ_k denotes the position of the current routing sensor. The optimal direction toward the next sensor can be chosen according to a weighted average of the gradient of the objective function and the direct connection between the current sensor and the optimum position:

$$\mathbf{d} = \beta \nabla J + (1 - \beta)(\zeta_o - \zeta_k),$$

where the parameter β can be chosen, for example, as a function of the inverse of the distance between the current and the optimum sensor positions: $\beta = \beta\left(\|\zeta_o - \zeta_k\|^{-1}\right)$. This routing mechanism allows adapting the routing direction to the distance from the optimum position. For small distances, it might be better to follow the gradient of the objective function for the steepest ascent, that is, the fastest information gain. For large distances from the optimum position where the objective function is flat and data is noisy, it is faster to directly go toward the maximum than to follow the gradient ascent.

In order to locally evaluate the objective function and its derivatives, the query needs to be transmitted together with information on the current belief state. This information should be a compact representation of the current estimate and its uncertainty and must provide complete information to incrementally update the belief state given local sensor measurements. For the earlier example of quantifying the information utility by the Mahalanobis distance, we need to transmit the triplet $\{\zeta_q, \hat{\mathbf{x}}, \hat{\Sigma}\}$ with the query, where ζ_q is the position of the querying sensor, $\hat{\mathbf{x}}$ is the current estimate of the target position, and $\hat{\Sigma}$ is the current estimate of the position uncertainty covariance.

The routing mechanism described earlier can be used to establish a routing path toward the potentially best sensor, along which the measurement from the sensor closest to the optimum position is shipped back. When global knowledge about optimum sensor positions is available, the routing path is optimal whereas information gathering may not be. In the case of local sensor knowledge, the path is only locally optimal because the routing algorithm is a greedy method. The estimate and the estimation uncertainty can be dynamically updated along the routing path. The measurement can also be shipped back to the query-originating node. Since the information utility objective function along the path is monotonically increasing, the information provided by subsequent sensors is getting incrementally better toward the global optimum. When the information is continuously shipped back to the querying sensor, the information arriving in sequential order provides an incremental improvement to the estimate. Once a predefined estimation accuracy is reached, the querying sensor can terminate the query even if it has not yet reached the optimum sensor position. Alternatively, instead of shipping information back to the querying sensor, the result could be read out from the network at the sensor where the information resides.

Simulation Experiments

The objective function used in the greedy routing experiments is chosen according to (5.1), with the information utility and cost terms defined, respectively, as:

$$\phi(\zeta_j, \hat{\mathbf{x}}, \Sigma) = -(\zeta_j - \hat{\mathbf{x}})^T \Sigma^{-1}(\zeta_j - \hat{\mathbf{x}}), \tag{5.6}$$

$$\psi(\zeta_j, \zeta_l) = (\zeta_j - \zeta_l)^T (\zeta_j - \zeta_l), \tag{5.7}$$

where ζ_l represents the position of the querying sensor l.

Figure 5.17 shows snapshots of numerical simulations of the greedy routing algorithm on networks of randomly placed sensors. To simplify the illustration, current target position, $\hat{\mathbf{x}}$, and its uncertainty, Σ, were arbitrarily chosen and remained fixed for the run—that is, no incremental update of the belief state was implemented. The value of the objective function across the sensor network is shown as a contour plot, with peaks of the objective function located at the center of ellipses. The circle indicated by a question mark (?) depicts the position of the querying sensor (query origin), and the circle indicated by T depicts the estimated target position, $\hat{\mathbf{x}}$. The current uncertainty in the position estimate, Σ, is depicted as a 90-percentile ellipsoid enclosing the position T.

The goal of the greedy routing algorithm is to guide the query as close as possible toward the maximum of the objective function, following the local gradients to maximize incremental information gain. While the case of trade-off parameter $\gamma = 1$ represents maximum information gain, ignoring the distance from the querying sensor (and hence the energy cost), the case $\gamma = 0$ minimizes the energy cost, ignoring the information gain. For other choices of $0 < \gamma < 1$, the composite objective function represents a trade-off between information gain and energy cost.

Figure 5.17 shows how variation of the trade-off parameter γ morphs the shape of the objective function. As γ decreases from 1 to 0, the peak location moves from being centered at the predicted target position ($\gamma = 1$) to the position of the querying sensor ($\gamma = 0$); at the same time, the contours change from being elongated, shaped according to the uncertainty ellipsoid represented by the estimated covariance Σ, toward isotropic. Another interesting aspect of the combined objective function is that the spatial position of its maximum does not shift linearly between the estimated target position T and the query origin ?, with varying γ. This can be observed in the case of $\gamma = 0.2$, where the maximum is located off the line connecting T and ?.

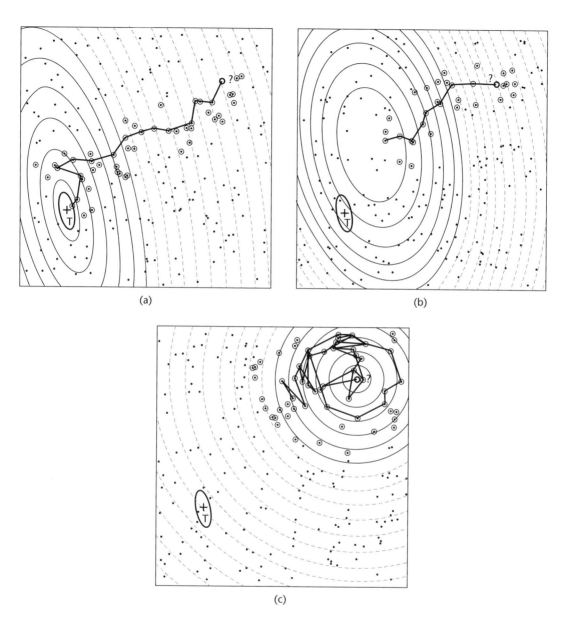

(a)

(b)

(c)

Figure 5.17 Locally optimal routing for $N = 200$ randomly placed sensors, with varying information versus cost trade-off parameter γ. From (a) to (c): $\gamma = 1$, $\gamma = 0.2$, $\gamma = 0.0$. For comparison, the position estimate of the target, T, and the position of the query origin, ?, are fixed in all examples (adapted from [43]).

In all three cases shown in Figure 5.17, the estimated target position and residual uncertainty are the same. Variations in shape and offset of the objective function are caused by variations of the trade-off parameter γ. In order to visualize how the query is routed toward the maximum of the objective function by local decisions, both the estimated position $\hat{x}$ as well as its uncertainty Σ are left unaltered during the routing. It is important to note that incremental belief update during the routing by in-network processing would dynamically change both the shape and the offset of the objective function according to the updated values of the estimated position $\hat{x}$ and its uncertainty Σ at every node along the routing path. As the updated values of $\hat{x}$ and Σ are passed on to the next node, all routing decisions are still made locally. Hence, the plotted objective function represents a snapshot of the objective function that an active routing node locally evaluates at a given time step, as opposed to the overlaid routing path which illustrates the temporal evolution of the multihop routing.

The small circles surrounding the dots along the routing path illustrate the subset of sensors the routing sensors (on the path) consider during sensor selection. Among these sensors, the ones that locally maximize the objective function have been selected as the successor routing nodes. The fraction of selected nodes among all nodes indicates the energy saved by using the greedy routing, as opposed to the total energy cost of flooding the network. The routing in Figure 5.17 can be terminated after reaching a spatial region where the residual uncertainty is below a preset threshold or the routing has reached a preset timeout that is passed along with the query.

5.4.2 Multistep Information-Directed Routing

The sensor selection in the previous routing problem is *greedy*, always selecting the best sensor given the current belief $p(\mathbf{x} \mid \{\mathbf{z}_i\}_{i \in U})$, and may get stuck at local maxima caused, for example, by network holes from the depletion of sensor nodes. Figure 5.18(a) provides a simple example. Here we use the inverse of Euclidean distance between a sensor and the target to measure the sensor's information contribution (assuming these information values are given by an "oracle"). The problem with greedy search exists regardless of the choice of information measure. Consider the case that the target moves from X to Y along a straight line [see Figure 5.18(a)]. Assume nodes A and B are equidistant from the target at any time. At time $t = 0$, suppose node A is the leader and can relay its target information to its neighbor B or C. If the selection criterion prefers a different node each time to increase diversity, then node B is chosen as the next leader. By the same criteria, B then relays back to A. The hand-offs continue back and forth between A and B, while the target moves away. The path never gets to nodes E, F, or G, who may become more informative as the target moves closer to Y. The culprit in this case is the "sensor hole" the target went through. The greedy algorithm fails due to its lack of knowledge beyond the immediate neighborhood. Recently, local routing algorithms such as GPSR [112] have been developed to traverse perimeters of network holes. However, they do not apply here since the routing destination is not always known a priori in our problem, and it is often impossible to tell if the routing is stuck at a local optimum without knowledge about the destination (e.g., compare the scenario in Figure 5.18(b) with that in Figure 5.18(a)).

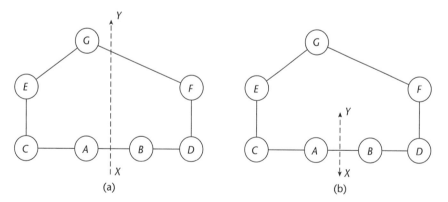

Figure 5.18 Routing in the presence of sensor holes. A through G are sensor nodes. All edges have unit communication cost. The dashed lines plot target trajectory. In (a), the target is moving from X to Y. In (b), the target is bouncing back and forth between X and Y (adapted from [145]).

To alleviate the problem of getting trapped at local optima, one may deploy a look-ahead strategy to extend the sensor selection over a finite look-ahead horizon. However, in general the information contribution of each sensor is state-dependent—that is, how much new information a sensor can bring depends on what is already known. This state-dependency property sets the information-directed routing problem apart from traditional routing problems. Standard shortest-path algorithms on graphs such as Dijkstra or Bellman-Ford are no longer applicable. Instead, the path-finding algorithm has to search through many possible paths, leading to combinatorial explosion.

To illustrate the state dependency in information aggregation, consider a simple sensor network example consisting of four sensors, A, B, C, and D, as shown in Figure 5.19. Belief about the target location is shown using grayscale grids (Figure 5.20). A brighter grid means that the target is more likely to be at the grid location.

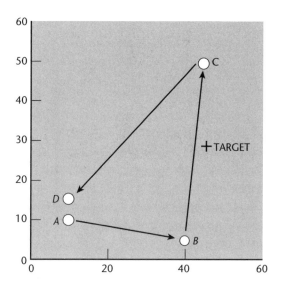

Figure 5.19 A sample sensor network layout: Sensors are marked by circles, with labels A, B, C, and D. Arrows represent the order in which sensor data are to be combined. The target is marked by + (adapted from [145]).

We assume a very weak initial belief, uniform over the entire sensor field, knowing only that the target is somewhere in the region. Figure 5.20(a)–(d) shows how information about the target location is updated as sensor data is combined in the order of $A \rightarrow B \rightarrow C \rightarrow D$.

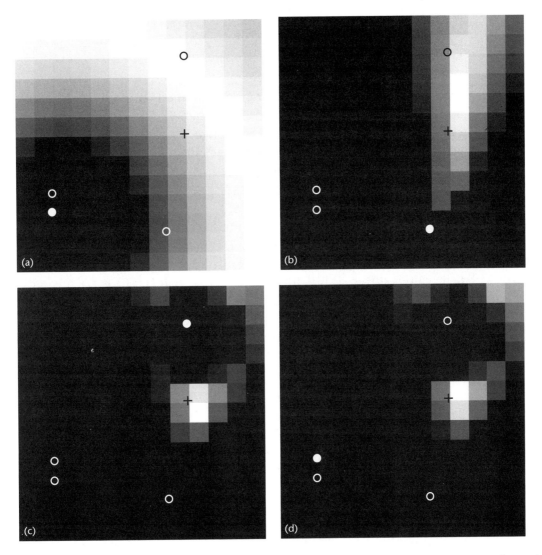

Figure 5.20 Progressive update of target position, as sensor data is aggregated along the path *ABCD*. Figures (a)–(d) plot the resulting belief after each update.

Table 5.1 Information aggregation in the sensor network pictured in Figure 5.19.

Order of traverse	Information*	MSE
Sensor A	0.67	11.15
Sensor B	1.33	10.02
Sensor C	1.01	9.00
Sensor D	0.07	8.54

*Information is measured using mutual information defined in (5.4).

At each step, the active sensor node, marked as a solid white dot, applies its measurement to update the belief. The localization accuracy is improved over time: the belief becomes more compact and its centroid moves closer to the true target location.

Table 5.1 lists the information contribution for each sensor, as the path $A \rightarrow B \rightarrow C \rightarrow D$ is traversed. The error in localization, measured as mean-squared error (MSE), generally decreases as more sensor measurements are incorporated. Note that sensors A and D are physically near each other, and their contributions toward the target localization should be similar. Despite such similarity, the information values differ significantly (0.67 for A and 0.07 for D). Visually, as can be observed from Figure 5.20, sensor A brings a significant change to the initial uniform belief. In contrast, sensor D hardly causes any change. The reason for the difference is that A applies to a uniform belief state, while D applies to a compact belief, as shown in Figure 5.20(c).

State dependency is an important property of sensor data aggregation, regardless of specific choices of information metrics. Sensor measurements are often correlated. Hence a sensor's measurement is not entirely new; it could be merely repeating what its neighbors have already reported. In our current example, sensor D is highly redundant with sensor A. Such redundancy shows up in the belief state and thus should be discounted. Because of this, the search cost function [say, defined as the path cost minus the information gain, similar to that in (5.1)] is not necessarily additive along a path.

To mitigate the combinatorial explosion problem, two strategies may be useful. We can restrict the search for optimal paths to a small region of the sensor network. Or we can apply heuristics to approximate the costs so that they can be treated as additive. The first technique we describe searches for a shortest path among the family of paths with fewer than M hops that produce maximum information aggregation. The look-ahead horizon M should be large enough and comparable to the diameter of sensor holes, yet not so large as to make the computational cost prohibitive. The information about network inhomogeneity may be discovered periodically and cached at each node (see, e.g., [102]). Such information will be helpful in selecting the value for M during the routing-path planning.

For scenarios such as the one in Figure 5.16(b), the goal is to route a query from the query proxy to the exit point and accumulate as much information as possible along the way, so that one can extract a good estimate about the target state at the exit node and yet keep the total communication cost close to some prespecified amount. When it is possible to estimate the cost to go, the A* heuristic search method may be used [120]. The basic A* is a best-first search, where the merit of a node is assessed as the sum of the actual cost g paid to reach it from the query proxy, and the estimated cost h to pay to get to the exit node (the "cost to go"). For real-time path-finding, we use a variation of the A* method, namely, the real-time A* (RTA*) search.[1] It restricts search to a small local region and makes real-time moves before the entire path is planned. Since only local information is used in the RTA* search, it can be implemented in a distributed fashion. Details of the above multistep search algorithms can be found in reference [145].

5.4.3   Sensor Group Management

In the scenarios we have considered so far (Figure 5.16), the physical phenomenon of interest is assumed to be stationary. In many applications, the physical phenomenon may be mobile, requiring the network to migrate the information according to the motion of the

---

1 This is closely related to the LRTA* algorithm of Section 3.4.4

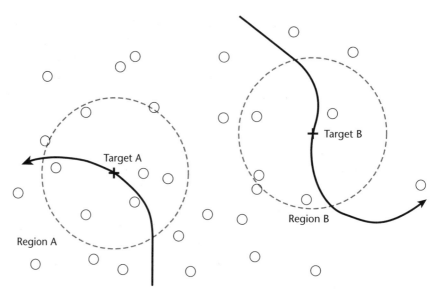

Figure 5.21 Geographically based collaborative groups. The small circles are sensor nodes. The nodes inside a specified geographical region (e.g., region A or B) form a collaborative group (adapted from [142]).

physical phenomenon for communication efficiency and scalability reasons.

In practical applications, the effect of a physical phenomena usually attenuates with distance, thus limiting the propagation of physical signals to geographical regions around the physical phenomenon. This gives rise to the idea of geographically based collaborative processing groups. In the target tracking problem, for example, one may organize the sensor network into geographical regions, as illustrated in Figure 5.21. Sensors in the region around target A are responsible for tracking A, those in the region around B for tracking B. Partitioning the network into local regions assigns network resources according to the potential contributions of individual sensors.

Furthermore, the physical phenomena being sensed change over time. This implies that the collaborative groups also need to be dynamic. As the target moves, the local region must move with it. Sensor nodes that were previously outside the group may join the group, and current members may drop out. This requires some method for managing the group membership dynamically.

Geographically based group initiation and management have to be achieved by a lightweight protocol distributed on all sensor nodes. The protocol needs to be powerful enough to handle complex situations, such as those where data from multiple leaders are contending for processing resources, and be robust enough to tolerate poor communication qualities, such as out-of-order delivery and lost or delayed packets. In addition, the propagation region of group management messages should be restrained to only the relevant nodes without flooding the entire network. This is not trivial, considering that the group membership is dynamic as the targets move and that the network is formed in an ad hoc way such that no nodes have the knowledge of the global network topology. The difficulties may be tackled via two techniques: (1) a leader-based tracking algorithm where at any time each group has a unique leader who knows the geographical region of the collaboration; and (2) recent advances in geographical routing (Section 3.4.4) that do not require the leader to know the exact members of its group.

Example: Information Migration in Tracking a Moving Target

How can the information utility measures be applied to a tracking problem such as the one described in Section 2.1? Assume a leader node (the solid dot in Figure 5.22) carries the current belief state. The leader chooses a sensor with good information in its neighborhood according to the information measure and then hands off the current belief to the chosen sensor (the new leader). As discussed earlier, the information-based approach to sensor querying and data routing selectively invokes sensors to minimize the number of sensing actions needed for a given accuracy and, hence, latency and energy usage.

To estimate the position of the target, a leader node updates the belief state information received from the previous leader with the current measurement information, using, for example, the sequential Bayesian method introduced in Section 2.2.3. For a moving target, a model on the target dynamics can be used to predict the

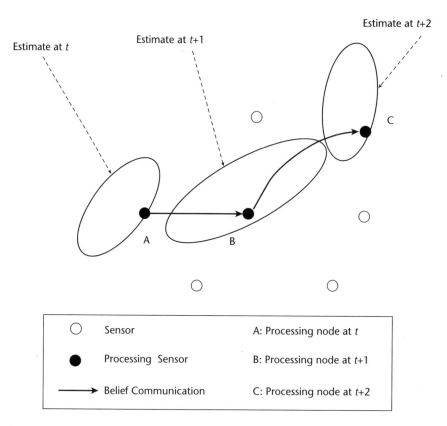

Figure 5.22 Information migration in tracking a moving target. As the target moves through the field of sensors, a subset of sensors are activated to carry the belief state. Each new sensor may be selected according to an information utility measure on the expected contribution of that sensor conditioned on the predicted location of the target (adapted from [41]).

position of the target at the next time step. This predicted target position and the associated uncertainty can be used to dynamically aim the information-directed query at future positions to optimally track the target.

Distributed Group Management

A collaborative group is a set of sensor nodes responsible for the creation and maintenance of a target's belief state over time, which we call a *track*. Effectively, these are sensors whose coverage overlaps with the state estimate of the track. When the target enters the sensor field or emits a signal for the first time, it is detected by a set of sensor nodes. Each individual sensor performs a local detection using a likelihood ratio test. Nodes with detections form a collaborative group and select a single leader—for instance, based on time of detection as discussed in Section 5.3.3. While we discuss the single-leader approach, it is also possible that a small number of nodes are elected to share the leadership. Figure 5.23 shows how the leader node maintains and migrates the collaborative processing group.

1. After the leader is elected, it initializes a belief state $p(\mathbf{x}|\mathbf{z})$ as a uniform disk R_{detect} centered at its own location [Figure 5.23(a)].

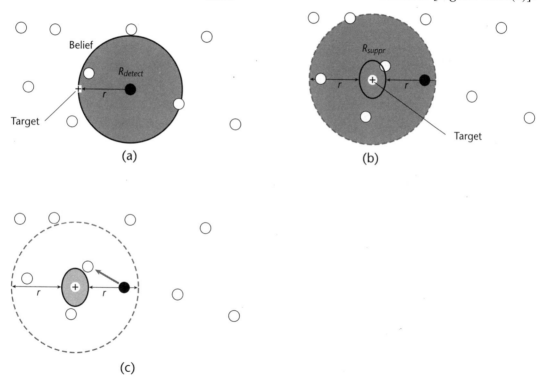

Figure 5.23 Distributed group maintenance: (a) An elected leader initializes a uniform belief over a region R_{detect}. (b) The leader estimates the target position and sends a suppression message to a region R_{suppr}. Nodes not receiving suppression time out to detection mode. (c) A new leader is selected using a sensor selection criterion. The current belief state is handed off to the new leader.

The disk contains the true target location with high probability. This belief provides a starting point for the subsequent tracking.

2. As the target moves, the sensors that did not previously detect may begin detecting. These sensors are potential sources of contention. The leader uses the uncertainty in track position estimate and maximum detection range to calculate a suppression region and informs all group members in the suppression region to stop detection [Figure 5.23(b)]. This reduces energy consumption of the other nodes and avoids further track initiation. The assumption is that there is only one target in the neighborhood.

3. Sensors are selected to acquire new measurements, using, for example, the sensor selection algorithm discussed earlier [Figure 5.23(c)]. As the belief state is refined by successive measurements, the group membership needs to be updated. This is accomplished by updating the suppression region using suppression and unsuppression messages to designated regions.

When the targets are far apart, their tracks are handled by multiple collaborative groups working in parallel. When targets cross, the position uncertainty regions for their tracks overlap and the collaborative groups for these tracks are no longer distinct. This can be detected when a leader node receives a suppression message from a node with a different (track) ID from its own. When two groups collide, the sensor measurements in the overlapping region can now be associated with either one of the two tracks.

Data association algorithms such as optimal assignment or multiple hypothesis processing can be used to resolve this ambiguity. For example, a simple track-merging approach is to keep the older track and drop the younger track. The two collaborative groups then merge into a single group. This approach works well if the two tracks were initiated from a single target. When the two tracks result from two targets, the merging operation will temporarily track the two targets as one. When they separate again, a new track corresponding to one of the targets will be reinitiated; however, the identities of the targets will be lost. Using an identity management algorithm, the ambiguities in the target identities after crossing tracks can be resolved using additional local evidence of the track identity and then propagate the information to other relevant tracks. Details of managing groups for multiple targets and their identities can be found in references [142, 210].

5.4.4 Case Study: Sensing Global Phenomena

We have been primarily concerned with sensing point targets so far. In some situations, we might encounter the problem of sensing a global phenomenon using local measurements only. In Section 5.3.4, we briefly described the problem of sensing a global relation among a set of objects. Another example of sensing global phenomena is determining and tracking the boundary of a large object moving over a sensor field, where each sensor only "sees" a portion of the boundary. One such application is tracking a moving chemical plume using airborne and ground-based chemical sensors.

How does the sensor tasking for these problems differ from what we have considered in sensing point targets? A primary challenge is to relate a local sensing action to the utility of determining the global property of the object(s) of interest. To address this challenge, we need to convert the global estimation and tracking problem into a local analysis, using, for example, the so-called *primal-dual transformation* [51, 140]. Just as a Fourier transform maps a global property of a signal such as periodicity in the time domain to a local feature in the frequency domain, the primal-dual transformation maps a line in the primal space into a point in the dual space, and vice versa (Figure 5.24). Using the primal-dual transformation, the shape of a target, when approximated as a polygonal object, can be tracked as a set of points in the dual space.

A useful consequence of this mapping is its use in tasking sensors to sense a global phenomenon such as the boundary of a moving half-plane shadow. As we noted earlier, a wireless sensor network is severely constrained by the on-board battery power. If a sensor only wakes up, senses, and communicates when it expects an event of interest, the power consumption of the network could

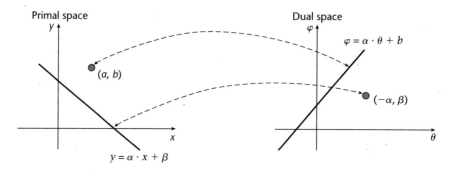

Figure 5.24 Primal-dual transformation. It is a one-one mapping where a point maps to a line and a line maps to a point (adapted from [140]).

be dramatically reduced. For the boundary tracking problem, the prediction of when a sensor needs to participate in a collaborative processing task can be made in the dual-space representation, where a boundary line L in the primal space is represented as a point l in the dual space, and sensors (points in the primal space) are represented as lines (Figure 5.25). Those lines that form a cell containing the point l correspond to sensors that are potentially relevant for the next sensing task. As the half-plane shadow moves in the physical space (i.e., primal space), the corresponding point in the dual space moves from cell to cell. When the point crosses the cell boundary,

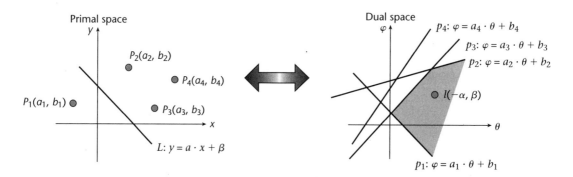

Figure 5.25 The prediction of which sensors will be relevant for sensing the target boundary L in the primal space is equivalent to the determination of lines that form the cell containing the corresponding l in the dual space arrangement (adapted from [140]).

a set of new sensors may become relevant to sensing the half-plane boundary and can be read off from the lines enclosing the current position in the dual space. The number of lines bounding a cell is 4 on the average. Thus the number of sensors that need be active at once is very small, no matter how many sensors are present in the field. This idea has been implemented and tested on a testbed of Berkeley wireless sensor motes (see reference [140]). An open problem that remains to be addressed is an effective decentralization of the computation in the dual-space containment test.

5.5 Summary

We have developed a number of important ideas for efficiently allocating the sensing, processing, and communication resources of a sensor network to monitoring and other application tasks. We have introduced the models of information utility and costs, as a basis for decentralized coordination and optimization *within* the network. The idea of information-driven sensor tasking, and its realization in IDSQ, is to base the sensor selection on the potential contribution of a sensor to the current estimation task while using a moderate amount of resources. Applying the idea to sensing stationary or moving physical phenomena, we developed a number of protocols, including the leader-based and moving center of aggregation. Moving beyond local greedy sensor selection, we introduced an information-driven routing to jointly optimize for routing and information aggregation, using a multistep look-ahead search. We also touched on the important topic of creating and managing collaborative processing sensor groups, which are common to a number of monitoring applications.

A number of key themes emerge from these discussions:

- Central to these ideas is the notion of information utility, and the associated costs of acquiring the information. In the resource-limited sensor networks, the appropriate balance between the information and the costs is of paramount concern, since unnecessary data collection or communication consumes precious bandwidth and energy and overload human attention.

- The information utility measures can take on many different forms, depending on the application requirement and context. Aside from information-theoretic considerations, we must carefully evaluate the computational complexity of applying the utility measures to sensor tasking, as inappropriate uses of information utility may consume intolerable amounts of resources and thereby diminish the benefit.

- As a sensor network's primary function is to collect information from a physical environment, we must rethink the role of routing in this context. As is becoming clear in the examples we examined, routing in a sensor network often does not just perform a pure information transport function. It must be co-optimized with the information aggregation or dissemination.

- A collaborative processing group is an important abstraction of physical sensors, since individual sensors are ephemeral and hence less important, and sensors collectively support a set of tasks. The challenge is to efficiently create, maintain, and migrate groups as tasks and physical environments change. A major benefit of establishing the collaborative group abstraction is in enabling the programming of sensor networks to move from addressing individual nodes to addressing collectives, a topic we discuss again in the context of platform issues and programming models in Chapter 7.

6

Sensor Network Databases

From a data storage point of view, one may think of a sensor network as a distributed database that collects physical measurements about the environment, indexes them, and then serves queries from users and other applications external to or from within the network. In this chapter, we study how sensor data is organized and stored after sensing actions, what user interfaces to the sensor database may look like, and how queries are processed and served in an efficient manner. The advantage of the database approach is that it provides a separation between the logical view (naming, access, operations) of the data held by the sensor network and the actual implementation of these operations on the physical network. Though any such abstraction comes with some loss of efficiency, this is compensated by a greatly increased ease of use. Diverse sensor network users and applications can focus on the logical structure of the queries they intend to pose and are relatively isolated from the details of physical storage and data networking on the volatile physical infrastructure of the network (sensors can fail, links come and go, and so on). Before discussing sensor-network-specific database issues, let us quickly review the structure of more traditional databases and certain relevant extensions.

In a classical database management system (DBMS) [78], data is stored in a centralized location. The structure and constraints of the data format, the so-called database *schema*, are typically defined or modified by a database administrator using a *data definition language* (DDL). Today most databases employ relational schemas and their variants [50], organizing data into tables whose rows are record tuples and whose columns are labeled by data attributes. A DDL compiler translates these definitions into *metadata*, a data structure describing the structure of the database data and the constraints they must satisfy, which is stored in permanent storage along with the actual data. In a typical database system, the *storage and buffer manager* directly controls storage devices, such as disks, and the flow of data between them and main memory. The database is updated through units of work named *transactions*. It is the job of the *transaction manager* to guarantee that transactions are executed atomically (that is, either in their entirety or not at all) and in apparent isolation from other transactions (concurrency control). This may include logging all transactions so that recovery is possible after disk or power failures.

The user or application is shielded from the details of how data is physically stored on hard disks and other devices. Instead, a user queries the database in a high-level logical *query language*, such as SQL [84]. A query is parsed by the *query processor* and translated into an optimized *execution plan*, which is then processed by the *execution engine* to answer the user query. Query processing may be aided by one or more *data indices* built in advance to facilitate such processing. In any database system, there are trade-offs between the speed of answering queries and the speed of performing database updates. A well-designed index structure can enable fast query response while still processing update transactions at moderate cost.

In distributed database systems [215, 121], data storage may be allocated among several geographically separated locations, connected by a communications network. The need for such distribution arises in many contexts—for instance, with businesses with geographically dispersed facilities. Data distribution along with data replication makes the entire system more robust to failures and can provide increased bandwidth and throughput, as well as greater data availability. A distributed database makes the job of the query processor significantly harder, however. Most query execution plans can be represented as trees where the nodes represent database operators (such as join, group-by, sort, or scan) and the edges correspond to producer-consumer relationships among operators. In a distributed setting, the nodes may live on physically distinct processors. Thus, in devising the best overall execution plan the query optimizer must consider the effects of simultaneous execution of multiple operators, as well as the communication cost of shipping data between processors over a network. Distributed databases can vary widely in regards to the number of physical nodes involved. A particular variant that has come into prominence recently is *peer-to-peer* (P2P) networks [9]. In such networks, active processors can number in the tens to hundreds of thousands and may come on the network and go off the network at arbitrary times. Though most such current systems (Napster, Gnutella, Kazaa) are limited to file sharing and thus have very primitive data models, the indexing and query processing issues they raise are at the research frontier, and they have many aspects in common with database problems for sensor networks.

Another recent trend in database systems is to consider systems for *data streams* [1, 163]. Such systems are aimed at handling long-running *continuous* queries, such as may arise in network or traffic monitoring, telecom call, stock market transaction, or web-log record summarization. The assumption here is that there is insufficient storage to hold all the data that has been (or can be) generated. The goal is to allow a user or application to query these data in a statistical or approximate form, by maintaining appropriate summaries.

In this chapter we first describe the various challenges that must be addressed in building sensor network databases (Section 6.1). We then focus on the logical structure of queries appropriate for sensed data (Sections 6.2 and 6.3), as well as the high-level database organization they imply (Section 6.4). Following that, we discuss techniques for query processing in sensor networks that benefit from in-network aggregation (Section 6.5). After a brief review of data-centric storage (Section 6.6), we go on to study data indices and range queries (Section 6.7). We end the chapter by discussing the topics of how to map onto the network topology hierarchically aggregated information, such as that used by various indices (Section 6.8), and of how to deal with temporal data (Section 6.9).

6.1 Sensor Database Challenges

Each sensor in a sensor network takes time-stamped measurements of physical phenomena such as heat, sound, light, pressure, or motion. Signal processing modules on a sensor may produce more abstract representations of the same data such as detection, classification, or tracking outputs. Additionally, a sensor contains descriptions of its characteristics, such as the location or type of the sensor. A sensor network database comprises all of the above data from every sensor. One approach for implementing such a database would be to transfer all these data to one or a small number of external warehouses, where a traditional DBMS system could be deployed. This, however, has several drawbacks, as we will see later on. The alternative is to store the data within the network itself and allow queries to be injected anywhere in the network.

It is important to realize that both the type of data obtained from sensors and the physical organization of a sensor network differ significantly from each of the database system types discussed above. At the most fundamental level, in a sensor network we have to conceptually view all the data the system might possibly acquire as a large virtual database, distinct from the data the system has actually sensed and/or stored. With controllable sensors, such as pan-tilt cameras, the system may have to choose which among two non-overlapping fields of view to actually sense so as to best serve the injected queries. Thus resource contention has to be considered even down to the level of what data can enter the physical database. Furthermore, while every database system has to grapple with the issue of what data to store, in sensor networks, more than in any other distributed database, the issue of *where* to store the data becomes of paramount importance since communication costs dominate the energy landscape.

At the physical level, there are two major distinguishing characteristics of sensor networks when it comes to database implementation. The first is that the network replaces the storage and buffer manager—data transfers are from data held in node memory as opposed to data blocks on disks. The second is that node memory is limited by cost and energy considerations, unlike disk storage that has become incredibly inexpensive. These differences generate several new challenges for sensor network databases:

- The system as a whole is highly volatile; nodes may be depleted, links may go down, and so on. Yet the database system has to hide all this from the end user or application and provide the illusion of a stable, robust environment in which queries run to completion.

- Relational tables are not static since new data is continuously being sensed. They are best regarded as append-only streams where certain useful reordering operations (such as sorting) are no longer available.

- The high energy cost of communication encourages in-networking processing during query execution. In general, query processing has to be closely coupled and co-optimized with the networking layer, as discussed in Chapter 3. Exactly how to best accomplish this is an active research area.

- Access to data may be hampered by arbitrarily long delays, and the rates at which input data arrives to a database operator (like a join) can be highly variable. As a result, it is not enough to make a query execution plan only once. Instead, the rates and availability of data have to be continuously monitored; operator location (in the network) and operator sequencing may need to be frequently updated to achieve optimal query execution.

- Limited storage on nodes and high communication costs imply that older data has to be discarded. The database system can try to maintain more high-level statistical summaries of the deleted information, so that queries about the past can still be answered in some form. Strategies for dealing with stale data are a topic of current investigation.

- Sensor tasking (Chapter 5) interacts in numerous ways with the sensor database system. A sensor serving multiple queries must decide how to best allocate its sensing actions so as to satisfy the queries. In doing so, the sensor may itself generate queries about data sensed in other parts of the network.

- Classical metrics of database system performance, such as throughput (number of queries executable per unit of time) or delay (maximum time elapsed for answering a query) may have to be adjusted in the sensor network context because of high variance in these quantities.

There are also significant differences between sensor network data and those of other databases at the logical level.

- Sensor network data consists of measurements from the physical world. Inherently such measurements include errors such as interference from other signals, device noise, and the like. As a result, exact queries do not make much sense in sensor networks. Instead, range queries (where we ask that certain attributes lie in certain intervals) and probabilistic or approximate queries are more appropriate.

- Additional operators have to be added to the query language to specify durations and sampling rates for the data to be acquired.

- While single-shot queries are possible and useful in sensor networks, we expect that a good fraction of the queries will be of the continuous, long-running type, such as monitoring the average temperature in a room. This makes sensor network databases more akin to data streams, and research on query languages in that area can be applicable here.

- Often sensor networks are deployed to monitor the environment and report exceptional conditions or other events of interest. Thus it is important to have operators for correlating sensor readings and comparing them with past statistics. The logical framework for such operations needs further development. The same applies to languages for describing event detections and action triggers.

Research in sensor network databases is relatively new and can benefit from current efforts in both data streams and P2P networks. It differs from the former in that sensor network nodes operate with limited energy, processing, and memory resources. It differs from the latter in that geographic location can be of great importance in deciding what information to store and what information to query for. Most physical phenomena exhibit coherence across space and time, and exploiting that is a major research issue for sensor network databases.

6.2 Querying the Physical Environment

As we mentioned, it is advantageous to express queries to a sensor network database at a logical, declarative level, using relational languages such as SQL. High-level interfaces allow nonexpert users to easily interact with the database. Moreover, formulating queries in a manner independent of the physical structure or organization of a sensor network shields users from the implementation details of the database system. For example, the actual layout and connectivity of a network may change over the time window within which a query is processed. It would be difficult for a nonexpert user to anticipate all the possible events and design the corresponding query execution plan.

Let us consider the following flood warning system as an example of SQL-style querying of sensor networks. A user from a state emergency management agency may send a query to the flood sensor database: "For the next three hours, retrieve every 10 minutes the maximum rainfall level in each county in Southern California, if it is greater than 3.0 inches." This is an example of a long-running, monitoring query, and can be expressed in the following SQL-like syntax:

```
SELECT max(Rainfall_Level), county
    FROM sensors
    WHERE state = California
    GROUP BY county
```

```
HAVING max(Rainfall_Level) > 3.0in
DURATION [now, now+180 min]
SAMPLING PERIOD 10 min
```

The only difference from the SQL syntax is the addition of the `duration` clause that specifies the period during which data is to be collected, and the `sampling period` clause, which specifies the frequency at which the query results are returned. The query is expressed over a single table comprising all the sensors in the network, each of which corresponds to a row in the table. It is assumed that the database schema is known at a fixed base station whenever we discuss SQL-style query processing. For a peer-to-peer system, where a query may originate from any node, the database schema will have to be broadcast to every node. The above query is an example of an aggregate query, meaning that the query result is computed by integrating data from a set of sensors. A query can also ask for relations or correlations among a set of events—for example, "Sound an alarm whenever two sensors within 10 meters of each other simultaneously detect an abnormal temperature"—or spawn subqueries—for example, "Obtain an ID whenever sensors in region *R* detect a person."

Long-running, *continuous* queries such as these report results over an extended time window. Additionally, one may ask *snapshot* queries concerning the data in the network at a given point in time—for example, "Retrieve the current rainfall level for all sensors in Southern California." There may also be *historical* queries that ask for aggregate information over historical data—for example, "Display the average rainfall level at all sensors for the last three months of the previous year."

To summarize, the queries on sensor networks may

- Aggregate data over a group of sensors or a time window.

- Contain conditions restricting the set of sensors from contributing data.

- Correlate data from different sensors.

- Trigger data collection or signal processing on sensor nodes.

- Spawn subqueries as necessary.

To efficiently support each type of these queries, we must understand how the data organization and query execution are to be meshed with the spatially distributed processing and communication of a sensor network.

A user interacting with a sensor database will typically issue a sequence of queries in order to obtain all the information he/she wants. For example, a user may query for aggregate information in a drill-down style, to locate an item of interest; or a user may try to correlate events at different times or locations in the network. Thus it is important to be able to use outputs from past queries as inputs to further commands. But it is also important to remember that not all users of a database system will be human operators. Programs running on the nodes themselves may generate queries in order to decide what sensing actions a node should take. For example, a node with a pan-tilt camera may need to know the direction of arrival of remote vehicles in order to properly orient itself to capture the vehicles.

For the approaches we consider in this chapter, we generally assume that node services will provide reasonably accurate time synchronization and node geographic location information to each node. Various node service techniques have been discussed in Chapter 4.

6.3 Query Interfaces

As an example of a sensor database query interface, we discuss the *Cougar* sensor network database system [20, 21] which maintains an SQL-type query interface for users at a front-end server connected to a sensor network. Distributed query execution is optimized for both resource usage and reaction time. The Cougar approach attempts to preserve as much as possible the abstraction and familiarity of a traditional data warehousing system, and to do so in an efficient manner.

6.3.1 Cougar Sensor Database and Abstract Data Types

Cougar represents each type of sensor in a network as an abstract data type (ADT), as in most modern object-relational databases [208]. An ADT provides controlled access to encapsulated data through a well-defined set of access functions. An ADT object in the Cougar database corresponds to a physical sensor in the real world. For example, the public interface of a seismic sensor ADT may comprise signal processing functions, such as short-time Fourier transform (STFT) and vibration signature analysis. In the Cougar data model, sensor measurements are represented as time series, where each measurement is associated with a time stamp; Cougar assumes that the nodes are time synchronized with one another reasonably well, so that there is no misalignment when multiple time series are aggregated. To account for the fact that a measurement for a sensor is not instantaneously available due to network delays, Cougar introduces virtual relations—relations that are not actually materialized as ordinary tables—in contrast to the base relations defined in the database schema. Whenever a signal processing function returns a value, a record is inserted into the virtual relation in an append-only manner, meaning that records are never updated or deleted. Virtual relations provide an effective way to treat ADT functions that do not return a value in a timely manner, as is the case for sensor networks.

It would be very expensive to transmit data from all the sensors to the front-end server where the query processing could be performed. Instead, Cougar considers distributed query processing in the network. For example, in the flood warning query above, one may push the selection (max(Rainfall_Level) > 3.0in) out to each sensor, so that only those that satisfy the condition return a virtual record (i.e., a Rainfall_Level measurement together with sensor ID and time stamp) to the front-end server. Overall, the Cougar system attempts to borrow as much as possible from the central warehousing approach and adapt it to a distributed sensor network.

6.3.2 Probabilistic Queries

Sensor data invariably contains measurement uncertainty due to device noise or environmental perturbations. Thus requests for exact sensor reading values need to be replaced with a formalism that allows expression of this inherent uncertainty. As a measurement typically is subject to many small and random perturbations, one way to model such an uncertainty in a database system is to introduce a Gaussian ADT (GADT), which models the uncertainty as a continuous probability distribution function (pdf) over possible measurement values [69]. A GADT can be efficiently represented simply by the mean μ and standard deviation σ of a Gaussian. Just like an ordinary ADT in a database, a GADT is a first-class object, with a set of well-defined functions such as Prob, Diff, and Conf, whose semantics are defined in [69]. Instead of giving a detailed mathematical definition for each function, we illustrate how one might pose probabilistic queries using GADTs to a sensor field containing temperature sensors. Using the function Prob (probability), one may ask the following query: "Retrieve from sensors all tuples whose temperature is within 0.5 degrees of 68 degrees, with at least 60 percent probability":

```
SELECT *
  FROM sensors
  WHERE Sensor.Temp.Prob([67.5,68.5] >= 0.6)
```

Another operation, the Diff (difference) primitive, is introduced to allow for probabilistic equality tests, so one can meaningfully compare different Gaussian variables (an exact match is an event of measure zero). This in turn permits the definition of relational joins, which is a key operator of the relational algebra. Unfortunately, computations of GADT functions involve integration of Gaussian functions over intervals or, equivalently, computing the well-known error function. This is an expensive computation to perform on nodes and makes indexing for GADT data especially demanding.

Range queries are another important class of sensor network queries. These will be discussed further in Section 6.7. Many other types of queries can be imagined according to the nature of the physical phenomenon being monitored. For example, a scalar field (temperature, pressure) over a sensor area can be viewed as a sampled elevation map, and thus tools from the geographical information systems (GIS) [148] community can be relevant. For instance, one might query for the iso-contours containing all regions hotter than a given threshold value.

6.4 High-Level Database Organization

To be able to efficiently process queries like those described earlier, the design of a sensor database must be coupled with the underlying routing infrastructure and with the application characteristics, such as its data generation and access patterns. To illustrate the trade-offs, let us examine two different ways in which data from a sensor network might be stored. In the centralized warehousing approach briefly mentioned earlier, each sensor forwards its data to a central server or warehouse connected to the network via an access point. Assuming reasonably uniform distribution of the nodes, we use the average routing distance (i.e., number of hops) from a node to the access point, to characterize the cost of communication—this would be $O(\sqrt{n})$, where n is the number of nodes in the network. The cost for the external-storage-based approach scales as $O(D \cdot \sqrt{n})$, where D is the total amount of data the sensors ship to the server. User queries do not incur additional sensor network communication cost since they are processed on the external server. However, there are distinct disadvantages to this centralized approach. The nodes near the access point become traffic hot spots and central points of failure; plus, they may be depleted of energy prematurely. This approach does not take advantage of in-network aggregation of data to reduce the communication load when only aggregate data needs to be reported. Also, sampling rates have to be set to be the highest that might be needed for any potential query, possibly further burdening the network with unnecessary traffic. Finally, as we have discussed earlier, customers of the data may be other applications running on nodes themselves in the network. For example, in a closed-loop flood control system, a flood detection may trigger actions for the local floor regulators. Sending data to and retrieving data from a remote central server may cause unnecessary delays.

An attractive alternative is to store the data *within* the network, using the so-called *in-network storage*. At the center of the design here is the appropriate choice of storage points for the data, which act as rendezvous points between data and queries (see Figure 6.1), so that the overhead to store and access the data is minimized and the overall load is balanced across the network. The communication cost for storing the data remains comparable to that of the warehousing approach. The query time, however, depends strongly on how the data is indexed. Flooding the network with a query can incur $O(n)$ communication cost and may be undesirable in many cases. In a data-centric storage system (to be discussed in Section 6.6), the communication overhead for a query can be reduced to $O(\sqrt{n})$ via geographic hashing methods. Most importantly, the in-network storage allows data to be aggregated before it is sent to an external query, takes advantage of locality of information for in-network queries, and, if designed carefully, load-balances the database costs across the nodes.

To compare different approaches and characterize the performance of a sensor network database system, one needs to define a

Figure 6.1 Rendezvous mechanism for a sensor network database. Data and queries meet at rendezvous points, which may be external to the network, local to nodes where the data originates, or somewhere else inside the network.

set of metrics. The following are adapted for sensor networks from metrics for general database systems:

- *Network usage* is characterized by

  - Total usage: The total number of packets sent in the network.
  - Hot spot usage: The maximal number of packets processed by any particular node. The hot spot usage impacts the overall network lifetime before partitioning.

- *Preprocessing time*: The time taken to construct an index.

- *Storage space requirement*: The storage for the data and index.

- *Query time*: The time taken to process a query, assemble an answer, and return this answer.

- *Throughput*: The average number of queries processed per unit of time.

- *Update and maintenance cost*: Costs such as processing sensor data insertions, deletions, or repairs when nodes fail.

To summarize, a sensor network database differs from a traditional centralized system in that the resources are severely constrained and querying processing is tightly coupled with networking and application semantics. When designing a sensor database, we desire the following properties:

1. *Persistence*: Data stored in the system must remain available to queries, despite sensor node failures and changes in the network topology.

2. *Consistency*: A query must be routed correctly to a node where the data are currently stored. If this node changes, queries and stored data must choose a new node consistently.

3. *Controlled access to data*: Different update operations must not undo one another's work, and queries must always see a valid state of the database.

4. *Scalability in network size*: As the number of nodes increases, the system's total storage capacity should increase, and the communication cost of the system should not grow unduly.

5. *Load balancing*: Storage should not unduly burden any one node. Nor should any node become a concentration point of communication.

6. *Topological generality*: The database architecture should work well on a broad range of network topologies.

6.5 In-Network Aggregation

Let us now see how in-network query processing can be used to provide substantial energy savings when serving aggregate queries. This savings is possible because combining data at intermediate nodes reduces the overall number of messages the network has to transmit, thus reducing communication and prolonging the lifetime of the network.

6.5.1 Query Propagation and Aggregation

Consider the following simple example, where an average reading is computed over a network of six nodes arranged in a three-level routing tree (Figure 6.2). In the server-based approach, where the aggregation occurs at an external server, each sensor sends its data directly to the server. This requires a total of 16 message transmissions. Alternatively, each sensor may compute a partial state record, consisting of {sum, count}, based on its data and that of its children, if there are any. This requires a total of only six message transmissions.[1]

In-network aggregation and query processing typically involve query propagation (or distribution) and data aggregation (or collection). To push a query to every node in a network, an efficient routing

---

1 We assume each partial state record can fit into a single message.

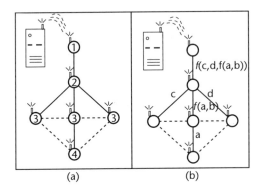

Figure 6.2   External server (a) versus in-network aggregation (b). In (a), the number at each node denotes the number of hops away from the external server (adapted from [149]).

structure has to be established. For example, a routing tree rooted at a base station could be used for this purpose. A query may be propagated through the routing structure using a broadcast mechanism (i.e., flooding the network). Or it may use multicast to reach only those nodes that may contribute to the query. For example, if the having-predicate specifies a geographic region, then portions of the routing structure that do not satisfy the having-predicates may be omitted from propagating the query.

Once a query has been distributed to all the nodes that satisfy the query conditions, data is then collected and aggregated within the network, utilizing the same routing structure. This gives rise to a number of important issues. Which aggregates can be computed piecewise and then combined incrementally? How should the activities of listening, processing, and transmitting be scheduled to minimize the communication overhead and reduce latency? How does the aggregation adapt to changing network structure and lossy communication? As discussed earlier, depending on whether a query is a snapshot, historical, or continuous, the aggregations are done over time windows of different sizes. For a snapshot query such as "report the current maximum temperature in a region," the aggregation needs to be completed within a small window of time within which the temperature is not likely to vary significantly. For other applications, the nodes must time-synchronize more accurately in order to correctly return the result. An example is the counting of moving vehicles on a highway. The vehicle movement may be such that different sensors may double count, or undercount, the vehicles if detections are not time stamped accurately. One major difference from a traditional database system is that for long-running queries in monitoring applications, a sensor network database returns a stream of aggregates, one per sampling period. A key challenge for in-network aggregation is the design of an optimal data aggregation schedule that is energy- and time-efficient.

6.5.2 TinyDB Query Processing

As an example, we consider TinyDB, a system designed to support in-network aggregate query processing [149]. TinyDB provides an SQL-style declarative query interface, and implements aggregation mechanisms that are sensitive to resource constraints and lossy communication. An example of TinyDB queries was given earlier in Section 6.2.

TinyDB supports five SQL operators: count, min, max, sum, average, and two extensions, median and histogram. Aggregation is implemented via a merging function f, an initializer i, and an evaluator e. The merging function f computes:

$$\langle z \rangle = f(\langle x \rangle, \langle y \rangle),$$

where $\langle x \rangle$ and $\langle y \rangle$ are multivalued partial state records. For example, for average, a partial state record $\langle S, C \rangle$ consists of sum and count of the sensor values it represents. The quantity $\langle z \rangle$ is the resulting partial state record, summarizing the data represented by $\langle x \rangle$ and $\langle y \rangle$:

$$f(\langle S_1, C_1 \rangle, \langle S_2, C_2 \rangle) = \langle S_1 + S_2, C_1 + C_2 \rangle.$$

The initializer i specifies the construction of a state record from a single sensor value. For average, this is $i(x) = \langle x, 1 \rangle$. Finally, the evaluator

e computes the value of an aggregate from the partial state record. For average, this is

$$e(\langle S, C \rangle) = S/C.$$

It is easy to show that other decomposable SQL queries such as max, min, count, and sum can also be expressed this way. For median, it can be shown that the size of the partial state record is proportional to the size of the data set it summarizes, while in histogram the size correlates with statistical properties, in this case the value distribution of the data set. These *holistic* or *content-sensitive* aggregates render the in-network processing less beneficial. In fact, simulation results from TinyDB support this observation (Figure 6.3). In general, we classify an aggregate according to the amount of state information required of each partial state record. Max, min, count, and sum are distributive, meaning the size of an intermediate state is the same as that of the final aggregate. Average is algebraic in that the partial state records are not themselves aggregates for the data set, but are of constant size. Count is unique, since the size is proportional to the size of distinct values in the data set. We have already seen that median is holistic, and histogram is content-sensitive. This classification is important because the performance of TinyDB is inversely related to the amount of intermediate state information required per aggregate.

6.5.3   Query Processing Scheduling and Optimization

In TinyDB, it is assumed that a query is issued from a server external to the sensor network. A routing tree rooted at the server distributes the query. The setup of the routing structure is discussed in Chapter 3. Here, we focus on how data is aggregated. TinyDB uses an epoch-based mechanism. Each epoch, or sampling period, is divided into time intervals. The number of intervals reflects the depth of the routing tree. Aggregation results are reported at the end of each sampling period, and the server receives a stream of aggregates, one per

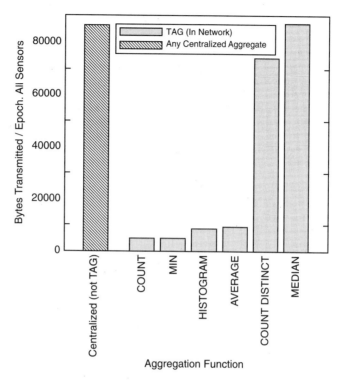

Figure 6.3   Benefits of in-network aggregation for different aggregation functions supported in TinyDB (adapted from [149]).

sampling period, over time. The choice of the sampling period is important—this period should be sufficiently large for data to travel from the deepest leaf nodes to the root of the tree.

Each node schedules its processing, listening, receiving, and transmitting periods according to its depth in the tree. When a node broadcasts a query, it specifies the time interval within which it expects to hear the result from its children. This interval is computed to end just a little before the time it is scheduled to send up its own result. During its scheduled interval, each node listens for the packets from the children, receives them, computes a new partial state record by combining its own data and the partial state records

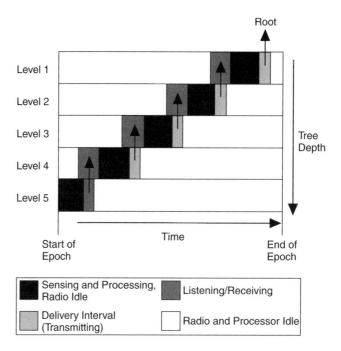

Figure 6.4   Schedule of node listening/receiving, sensing/processing, and transmitting in the multistage in-network aggregation in TinyDB (adapted from [149]).

from its children, and sends the result up the tree to its parent. This way, each node only needs to power up during its scheduled interval and so saves energy. Figure 6.4 shows an example of such a schedule. Notice that the listening interval of each parent is slightly larger than those of the children and includes the transmission interval of the children. This is to tolerate small inaccuracies in the node time synchronization. To increase the throughput, the aggregation operations at various depths of the tree may be pipelined, so that at each subinterval an aggregate is computed. TinyDB queries may include grouping predicates, such as (GROUP BY county) in the query example shown earlier. When grouping predicates are present, each partial record is tagged with a group ID, and only those with the same group ID may be aggregated at intermediate nodes.

One may implement additional optimizations over the basic structure of TinyDB. For example, a node may decide whether to contribute its data by snooping on others' packets. It may report data only if it has changed over previously reported data, or more generally, using a hypothesis testing to see if its data is going to affect the final value of the aggregate. A more important issue is to ensure robustness to node or link failures. Losing a parent node may orphan an entire subtree. Each child node has to periodically rediscover parents to make sure it is connected. TinyDB also considers providing redundancy by duplicating parent nodes for each child and by caching data over a past window of time at each node in the event the connections to the children are lost.

TinyDB builds a static tree for aggregation. However, as network conditions change or the rate at which data is generated varies, it is desirable to adapt the aggregation tree, so as to optimize the performance and resource usage of the database. This problem is studied under the name *adaptive query operator placement* in the literature [19]. The need for adaptive aggregation is illustrated in Figure 6.5. A user (at the sink) is notified whenever there are correlations between the detections from two regions, regions *A* and *B*. Assume each region elects a representative node to produce the detections for the region. Ideally, the correlation operator should be placed to minimize the

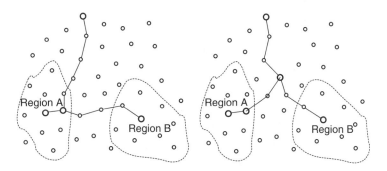

Figure 6.5   Scenarios of query operator placement. (a) Unequal number of detections from each region, and with low correlation. (b) Equal number of detections from each region and with some correlation (adapted from [19]).

overall communication overhead. In Figure 6.5(a), assume the detections from the two regions are not correlated, and region A generates more detections than region B. It is clear, then, that the correlation operator should be placed somewhere on the shortest path between the two nodes generating the detections for the regions and should be closer to region A. This is due to the fact that the correlation operator produces little data that needs to be shipped to the sink. On the other hand, if the detections from the two regions are somewhat correlated and are generated at approximately the same rate, then the correlation operator should be placed closer to the sink and at an equal distance to the two regions, as shown in Figure 6.5(b).

One can think of the problem of optimal operator placement as a task assignment problem, in which a set of tasks needs to be assigned to a network of processors. In our problem, the task is the query tree where nodes are the operators and edges are data dependencies among the operators. The network of processors are the sensor network graph. We would like to assign the operators to sensor nodes so as to minimize the communication cost. In general, this task assignment problem is NP-complete. In the case of the queries we are interested in, however, where the set of tasks is tree structured, polynomial algorithms solving the problem exist. These algorithms are still too complex for in-network implementation; a local optimization strategy [19] has been proposed that progressively moves operators toward locally optimal placement. The technique is rooted in a physics-based analogy. The cost constraints (for example, the data transfer cost between the nodes) can be thought of as elastic bands exerting forces on pairs of nodes. Operators will move from node to node toward an equilibrium position where this artificial system energy is minimal, corresponding to a locally least-cost position of the operators. A nice property of this technique is that the placement of the operators can be continuously adjusted as external conditions change.

Continuous adaptation of the operator ordering needed to execute a query has been considered in the Eddies system [10].

6.6 Data-Centric Storage

A tree-based query propagation mechanism such as the one used by TinyDB is appropriate for a server-based application. To support data access when queries may originate from multiple and arbitrary nodes within a network, however, more flexible storage and access mechanisms are needed. One may think of using multiple trees, one for each querying node, and overlaying them to share the same physical network. But this requires far more sophisticated scheduling than that used in TinyDB. Furthermore, it is still server-centric and requires flooding the network with the queries. Data-centric storage (DCS) is a method proposed to support queries from any node in the network by providing a rendezvous mechanism for data and queries that avoids flooding the entire network.

Just as in data-centric routing (such as directed diffusion, discussed in Section 3.5.1), DCS names and stores data by its (physical) attributes external to the network, rather than by network properties such as addresses of nodes where the data was generated. Queries retrieve data also based on these (physical) attributes. At the center of a DCS system are rendezvous points, where data and queries meet. A user queries the network by data attributes, or keys, and DCS provides a mechanism for translating the attributes into a node location or ID where the data is stored. Compared to local data storage on the node where the data was generated, or external warehouse-based storage, DCS distributes the storage load across the entire network. Such load balancing is effective when most queries can be resolved by utilizing only a small fraction of the data stored in the network.

An instance of DCS is the *geographic hash table* (GHT) [192, 191], as discussed in Section 3.5.3. To review, in GHT, the translation from node attribute(s) to storage location is accomplished by a hash function, which attempts to distribute data evenly across the network. GHT assumes each node knows its geographic location, using, for example, a GPS or other node location service. A data object is associated with a key, and each node in the system is responsible for storing a certain range of keys. In fact, any key may be used, as long as it can uniquely identify this data in a query. The queries considered in GHT are limited to point queries in which an exact match with data is desired. Techniques for range queries will be discussed in the next section.

GHT hashes keys into geographic coordinates and stores a key-value pair at the sensor node geographically nearest to the hash of its key. The system replicates stored data locally to ensure persistence, in case nodes fail. A geographic routing algorithm allows any node in the system to locate the storage node for an arbitrary key. GHT uses a modified GPSR, a routing algorithm discussed in Section 3.5.3, as the delivery routing mechanism. This enables nodes to *put* and *get* data based on their keys, thereby supporting a hash-table-like interface. As already discussed, data consistency and persistence are achieved by replicating data at nodes around the location to which the key hashes; furthermore, GHT ensures that one node is chosen consistently as the home node for that key. If too many events with the same key are detected, that key's home node could become a hot spot, for both communication and storage. The hot spot may be avoided using a replication scheme in which the key space is hierarchically decomposed, as shown in Figure 6.6. In this structured replication, the home node, called the *root* for that key, is replicated at $4^d - 1$ mirror sites. The figure shows replication for $d = 2$. To query for an event, the query is first routed to the root for the key, then from the root to the three level-1 mirrors. Each of these recursively forwards the query to the three level-2 mirrors associated with it, and so on. The replication scheme reduces the storage cost per node, at the expense of increased query cost.

There are several ways the original GHT may be extended. In many applications, data collected by a node may only be needed by other nodes that are nearby the data-originating node. In order to reduce unnecessary network traffic, this calls for hashing to locations that respect geographic proximity. Second, it is often more beneficial to hash to regions rather than to locations, for reasons such as avoiding hot spots and increasing robustness. The structured replication is one way to do this. Additionally, hashing to a region will couple the data index with a spatial index so that data may be efficiently searched

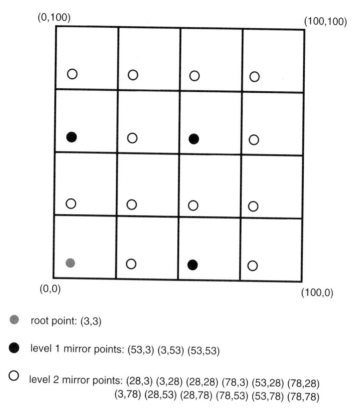

Figure 6.6 Structured replication in GHT (adapted from [192]).

for using a spatial indexing data structure. We will discuss all these topics later in this chapter.

DCS may be regarded as a variant of publish-and-subscribe [100], except the event broker in the publish-and-subscribe model now serves as the data storage and rendezvous point. In publish-and-subscribe, data may be pushed or pulled, depending on the relative frequencies and cost of push and pull. Similar trade-offs exist here. If the frequency of event generation is high, then pushing data to arbitrary rendezvous points may be too expensive, and local storage is preferred. A future research topic is to design adaptive DCS schemes that balance push and pull mechanisms based on relative data and query frequencies.

6.7 Data Indices and Range Queries

A type of query that is especially appropriate for sensor network databases is a *range query*. In such queries a certain "range" is specified for a number of attributes of interest on the data sought. For example, in our Atlantic petrel bird example from Chapter 1, a biologist may want to issue the following query: "Retrieve all petrel nesting events whose temperature lies between 50 and 60 degrees, and whose light readings is between 5 and 10 candelas per meter squared." In general, sensor network data is multi-attribute. In the petrel example, each nesting event may have associated temperature, humidity, light, wind speed, and time-of-day readings. In this example, each attribute can be parameterized by a single scalar value, and a range along one of the attributes corresponds to an interval, as in the example given. Implicit in our biologist's query is the fact that there are no restrictions along each of the attributes not mentioned. The ability to specify ranges also overcomes issues about possible errors and inaccuracies in the readings stored. Note that some attributes may be parameterizable only by higher-dimensional spaces, as for example, the three-dimensional orientation of an aircraft.

It is difficult to serve a range query well using only the techniques mentioned so far. A TinyDB-style aggregation tree can be built, but it would require flooding the entire network with the query each time (though this would still require less communication than a central warehousing approach). GHT, on the other hand, is aimed at point queries or exact matches, so it is not well suited for queries involving data ranges. Especially in a setting where the rate of queries on a sensor net database is high compared with the rate of updates, it makes sense to build auxiliary data structures, called *indices* in the database lingo, that facilitate and speed up the execution of the query. In our petrel bird database, it makes sense to build such an index for all the past events the system maintains, as that set is now static. The index can be rebuilt once a day to accommodate new detections.

In general, the complexity of answering a query in a sensor network will be a function of the size of the data in the network (say n, the number of records stored in all nodes), and the number of records returned (say k, the number of detections meeting the range constraints). Even if $k = 0$, there is an overhead in processing the query so as to verify that no matching events exist. The two main classical measures of the quality of an index are the speed at which it allows queries to be processed and its size. Clearly, there is a trade-off between these two measures: the more data the index stores, the faster the query processing can be. A third measure of index quality is the preprocessing cost of building the index and the cost of updating the index as changes are entered in the database. In our distributed database setting, communication costs also enter the picture. In particular, we have to decide both what the index will be and where its data will be stored in the network.

6.7.1 One-Dimensional Indices

The bulk of indexing research in the database community has dealt with one-dimensional indices [78], that is, indices for data that can be parameterized by a single value, and a variety of index structures have been developed for this setting based on B-trees, hash tables, and other structures. Most of these do not map directly to a networked implementation on small devices, nor do they deal with the special needs of indexing for range queries. Indices for range queries have been developed by the computational geometry community, where the need for such queries frequently arises [51]. In the context of range searching, a key idea is that of pre-storing the answers to certain special queries and then delivering the answer to an arbitrary range query by combining an appropriate subset of the pre-stored answers [154, 4]. The particular subsets of data forming these pre-stored answers are referred to as the *canonical subsets*. Again, there is a trade-off between the number of pre-stored answers and the speed of query execution. At one end, if nothing is stored, every record has to be queried. At the other end, if the answer to every possible query has been already computed, the cost of answering any query is simply to deliver this precomputed answer.

To illustrate the idea of canonical subsets, consider the following simple scenario. A number of traffic sensors are positioned along

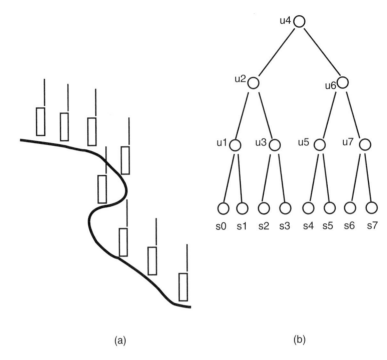

(a)   (b)

Figure 6.7   Canonical subsets of sensors along a road.

a road; see Figure 6.7(a). Over the course of a day, these sensors measure the traffic past them and accumulate traffic counts every minute. The local county road department wishes to decide where the road has to be widened to accommodate increased traffic. Civil engineers accessing this database will issue commands asking for traffic data along various contiguous segments of the road. The data to be returned is the per-minute counts averaged over all sensors in the segment.

This is a one-dimensional range search problem. It can, of course, be solved by building an aggregation tree for all the sensors for each queried segment of the road. However, multiple queries will redo a lot of the same aggregations, and repeatedly transmitting detailed traffic data can be expensive. Thus we proceed as follows. Conceptually, we build a balanced binary tree on the eight sensor nodes

$s_0, s_1, \ldots, s_7$, as shown in Figure 6.7(b). An internal node in the tree aggregates information from all its descendant sensors in the tree. For example, node u_2 aggregates data from sensors s_0, s_1, s_2, and s_3. We map logical node u_i to physical sensor node s_{i-1}, for $i = 1, 2, \ldots, 7$. In this example, the canonical subsets whose aggregations have been pre-stored are as follows ($\oplus$ denotes the aggregation operator):

u_1	$s_0 \oplus s_1$
u_2	$s_0 \oplus s_1 \oplus s_2 \oplus s_3$
u_3	$s_2 \oplus s_3$
u_4	$s_0 \oplus s_1 \oplus s_2 \oplus s_3 \oplus s_4 \oplus s_5 \oplus s_6 \oplus s_7$
u_5	$s_4 \oplus s_5$
u_6	$s_4 \oplus s_5 \oplus s_6 \oplus s_7$
u_7	$s_6 \oplus s_7$

Now if we query, say, for the segment of the road between s_0 and s_4, the desired answer can be obtained by just combining that stored in u_2 (physically in s_1) with that in s_4. In general, any segment query can be answered by combining the answer from all maximal nodes whose leaf descendants lie wholly in the segment (range); by maximal here we mean that the parent of that node does not satisfy that condition. In general, when we have n sensors along the road, this method will store $O(n)$ canonical subsets so that the answer to any segment query can be composed out of $O(\log n)$ of them: these are the pre-stored answers in the roots of the subtrees hanging to the right of the path from the left endpoint of the segment to the root and to the left of the path from the right endpoint to the root, until reaching the least common ancestor of the two nodes containing the endpoints.

As this example makes clear, partial data aggregation is a key feature of any indexing scheme for range searching. A difficulty with such hierarchical structures in the sensor network context, however, is that nodes higher up in the hierarchy are used more frequently in processing queries than other nodes and thus may be depleted prematurely. We will see a number of approaches addressing this problem in the following.

6.7.2 Multidimensional Indices for Orthogonal Range Searching

Sensor data is rarely indexed by a single attribute alone, such as road position in the earlier example. Even queries based on sensor location only will require two attributes in most settings, as sensors are typically deployed over a two-dimensional domain. When we have a set of attributes each parameterized by a scalar value and query with a range (interval) along a subset of the parameters, then we speak of *orthogonal range searching*. This name is apt because we can think of each attribute as a dimension in a high-dimensional space; the Cartesian product of a bunch of intervals along each of the axes is a rectangular parallelopiped aligned with the axes (think of dimensions along which no range is given as having an infinite range covering the entire axis). For example, when we parameterize locations by x and y coordinates in the plane, an orthogonal range query refers to all sensors inside some axis-aligned rectangle. Or, if we go back to our petrel-nesting example, the query

```
SELECT *
  FROM Nesting_Events
  WHERE Temperature >= 50 AND Temperature <= 60 AND
    Light >= 5 AND Light <= 10
```

can be visualized as in Figure 6.8.

Building indices for multidimensional data is a lot more challenging than the one-dimensional case. A direct use of one-dimensional techniques can be very inefficient. For example, we could build two separate one-dimensional indices for the petrel-nesting example, one on temperature and one on light. We can then use one-dimensional range searching to retrieve all nesting events where the temperature is in the desired range and, separately, all nesting events where the light is in the desired interval. Finally, these two sets of records can be intersected to produce the final answer. However, as Figure 6.8 makes clear visually, in doing so we will be retrieving many more records than necessary. The same problem arises with techniques that try to fold the two attributes into a single key. A one-dimensional space

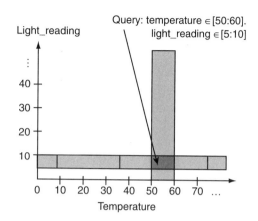

Figure 6.8   An example of an orthogonal range query.

cannot have the same neighborhood structure as a two-dimensional space, and again there is a lot of unnecessarily retrieved data that has to be filtered out.

A variety of true multidimensional indexing techniques have been developed based on both hashing/partitioning schemes and tree structures [78]. Grid files and partitioned hashing are examples of the former type. With one exception to be mentioned later, hashing schemes in general do not accommodate range searching well, as hash functions aim to spread the data around and not to preserve locality. Tree-based index structures include multilevel indices, *k-d* trees, quad-trees (and their higher-dimensional analogs), and *R* trees. All of these can be adapted for multidimensional range searching in the orthogonal case.

In each case, the goal of the structure is to quickly filter out those portions of the database that cannot possibly contain records relevant to a particular query. This is done in a top-down hierarchical fashion; thus the filtering that happens early on is the most effective. Let us illustrate this in a two-dimensional setting. Since quad-trees are likely to be already familiar to most readers, we use *k-d* trees as an example. Consider a small sample of nesting events in our petrel scenario, represented by points in a two-dimensional plane

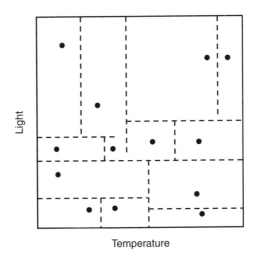

Figure 6.9   A *k-d* tree partitions a plane into rectangles.

parameterized by their temperature and light values, as in Figure 6.9. Suppose our query is to simply count the number of nesting events satisfying given temperature and light ranges. In our sensors on a road example, we built a perfectly balanced binary tree on the sensor nodes. One way of doing this is to repeatedly partition the nodes into equal-sized groups in their order along the road. We follow the same approach here, except we alternate cuts along the two dimensions. So (say) we first partition the events into two almost-equal-size groups by partitioning along the light axis. This generates two subgroups that are then partitioned, each separately, by cuts along the temperature axis. We proceed in this way, alternating between light cuts and temperature cuts across the levels, until each region contains just one (or a small number) of events.

Note that each node in this *k-d* tree represents a rectangular region in the temperature-light plane and can record the count of all the events stored inside it (we do not discuss now where to store this information in the network). As in our one-dimensional example, the nodes of this tree define the canonical subsets of the original events over which aggregated information is pre-stored. When a query is now given, it represents another rectangle Q in the temperature-light plane. We can drill down the k-d tree with this rectangle Q. Whenever we reach a node whose corresponding rectangle is disjoint from Q, we can just stop propagating. When we encounter a node whose corresponding rectangle is fully contained in Q, we can incorporate its count into a running total for the events of interest. Only when both of these conditions fail do we need to expand a node and continue drilling on its children. It is obvious that the storage cost for storing m events into such a tree is $O(m)$; it can also be shown that the worst-case overhead cost in processing a query is $O(\sqrt{m})$ [in d dimensions this becomes $O(m^{1-1/d})$] [51].

R trees [90] are another hierarchical data structure based on axis-aligned rectangles. However, they do not require that the rectangles of the children of the node partition the rectangle of the parent, as in k-d trees. The children rectangles may overlap and need not cover the entire rectangle of the parent, as long as they cover all the records that the parent covers. In general, the goal is to minimize the overlap among the children rectangles. The overlap may lead to some duplicate work, but the flexibility it provides has proved very useful.

Finally, multilevel indices can be composed out of one-dimensional indices. Again in our petrel example, one can build a one-dimensional index based (say) on the temperature attribute. Now, for each canonical piece of that index, a secondary index can be built based on the light attribute. This will increase the storage in the index to $O(n \log n)$ (each record appears in $\log n$ canonical pieces in the first index), but the query overhead can now be reduced to $O(\log^2 n)$ and then further to $O(\log n)$ with some additional effort [this becomes $O(\log^{d-1} n)$ in a d-dimensional attribute space]. More details can be found in [51].

6.7.3 Nonorthogonal Range Searching

Forcing all attributes to be one-dimensional is sometimes overly constraining. For instance, even if we want to search based on sensor location only, viewing the sensor location as two independent attributes (x- and y-coordinates) is unnatural—not every region of interest can be described well as a rectangle aligned with the axes. Techniques have been developed that deal with more general query shapes over attributes of dimension higher than one. However, they tend to be significantly more complex.

As an example, suppose we have n sensors fairly evenly spread over a two-dimensional area. Suppose further that these sensors are able to detect, localize, and count the number of targets in a small region around themselves, either in isolation or in a lightweight collaboration with their neighbors. This local region diameter is assumed to be comparable to the sensor spacing. Our global goal is to count the total number of targets present in certain types of simple nonorthogonal geometric areas—and for the sake of this example we assume that these areas must be halfspaces. The target count can be obtained easily in $O(n)$ time by interrogating and aggregating data from each of the sensors via flooding.

However, if we are willing to precompute and store some partial results, then we can do significantly better. Suppose we put a $\sqrt{n} \times \sqrt{n}$ grid over the sensor network so that in each of its cells there is at least one sensor whose sensing region contains the cell, as in Figure 6.10. We now activate the sensors and count the number of targets within each of the grid cells; furthermore, we propagate these counts along rows of the grid, so that in the end, we know not only the total target count in the cell itself, but also the aggregate count for the cells to its left in its row, and to its right in its row.

Consider now a halfspace query, as shown in Figure 6.10. Note that, except for at most $2\sqrt{n}$ grid cells intersected by the halfspace edge, all other cells are either fully contained in the range or are fully outside the range. We can easily compute the total number of targets in the halfspace by aggregating target counts from the partially covered cells, plus the relevant pre-stored totals for each row from their left or right neighbors, as appropriate. Thus, by using storage only for local targets and two counts in each node, we are able to answer the halfspace target counting query in $O(\sqrt{n})$ time. Note that this applies to *any* halfspace query: our preprocessing has exploited knowledge of the general shape of the query (halfspace) but *not* of its exact parameters. By using more sophisticated geometric methods,

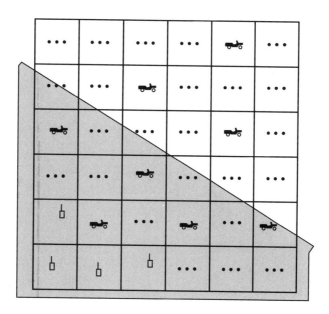

Figure 6.10 Counting the targets in a halfspace (adapted from [88]).

the same can be done for target counting in areas that have other simple geometric shapes; furthermore, the assumption about even distribution of the sensors can be dropped—as long as the union of the sensed regions covers the domain of interest. More details about this type of more advanced geometric searching can be found in [154, 4].

6.8 Distributed Hierarchical Aggregation

The range search data structures described in the previous section largely employ hierarchical aggregation of data to allow efficient query execution with a modest storage cost. Even without range searching in the picture, hierarchical aggregation is one of the main tools by which we can impose structure and organize the data obtained by a sensor network. In a distributed network setting, however, these tree structures must be overlaid on the network structure of the nodes. Except in a few simple examples we presented (the road traffic statistics and target counting problems), it is not clear how this mapping should be accomplished. Furthermore, as we briefly mentioned, the root and nodes near the root may well become traffic bottlenecks. Unlike a centralized system where data is inserted from the root, in sensor networks data is generated from the sensors that represent leaf nodes in the tree. Any new data generated from a sensor may have to be propagated all the way up to the root in order to update the index. Likewise, all access to sensor data will have to go through the root and drill down on the search tree. An important consideration in designing a distributed index data structure is load-balancing the communication, processing, and storage across the nodes. At the center of such designs is a routing infrastructure, or overlay, that optimizes and balances indexing traffic over the network topology. Because of frequent failures of nodes and links, robustness considerations necessitate replication mechanisms that provide some degree of data persistence.

This topic of appropriate spatial embeddings for hierarchical structures is of central importance in sensor network databases and is only now starting to receive the attention it deserves. In the following we describe some current efforts in this direction.

6.8.1   Multiresolution Summarization

Range searching indices provide a hierarchical summarization of records or events in the database according to simple attributes of interest. These summarizations are directly aimed at serving the queries allowed. More generally, we may want to obtain statistical summaries of such records or events at multiple spatial or temporal scales that can be examined by a human user as well as by programs. For example, an environmentalist may be interested in the concentration of contaminant flows over a county. Instead of flooding the entire county with the query, the user may try to first determine which regions might contain such phenomena, and then drill down on subregions of those regions. As another example, a traffic analyst attempting to set lights to improve traffic flow may be interested in vehicle speed data in a city, through a neighborhood, at an intersection, or over the past week, past day, or past hour. Because of storage and communication constraints, it is beneficial to summarize the spatio-temporal data in the network at multiple resolutions, so that the summaries can be easily queried, finer details can be found by drilling down the summarization tree, correlations among the data can be detected, and so on.

Wavelet transforms provide one way to compress and summarize information for both temporal and spatial signals and are widely used in signal and image processing. Wavelet compression extracts key features of the signal, such as discontinuities or long-term trends, that can be key in understanding the phenomena of interest. Readers are referred to [151] for a catalog of available wavelets and transforms such as sub-band coding. To summarize the data over a spatial domain, that domain has to be decomposed into pieces in a hierarchical fashion. A wavelet compression scheme separately computes the wavelet transforms for each of the subareas first, and then merges the resulting wavelet coefficients to form the wavelet transform for the overall region. To maintain storage and communication efficiency, at some level of the hierarchy, the higher-level transform is thresholded so that the size of the transform does not grow as it moves up the summarization hierarchy where more data is represented. The storage efficiency is obtained at the cost of losing some detail.

To support the spatial summarization in a sensor network, one needs to develop a data structure and a routing overlay. A quad-tree, for example, may be used for such purposes. A quad-tree recursively decomposes a space into quadrants. For each of the spatial cells in the decomposition tree, a sensor node can be chosen by a geographically constrained hash, as in GHT, to act as a representative. Routing between different cells can be done using GPSR. To avoid hot spots near the root of the tree, replication may be necessary. Alternatively, representatives for the higher tree nodes may be moved every so often to even out the load.

To query the multi-resolution summary, a query will start at the root of the summarization tree. If the summary stored at the root does not provide enough detail to form an answer, then the query will recursively descend down the tree, until the resolution is sufficient. At the heart of the drill-down querying is an indexing structure; at each node the query is matched against summaries from each of the children and, just like in range searching, nonmatching branches of the tree are pruned off.

These ideas have been implemented in a number of systems. DIMENSIONS is a system that provides multi-resolution storage and search for sensor data [72, 73]. That system also deals with temporal summarization, a topic we take up in Section 6.9. TinyDB has also been extended to provide a wavelet-based summarization [96].

6.8.2 Partitioning the Summaries

If we can partition aggregated data in a meaningful way, then we can distribute that data over several nodes in the network and thus lessen the load on nodes near the hierarchy root. A system that has developed this approach is DIFS (DIFS stands for "distributed index for features in sensor networks") [83].

DIFS considers a one-dimensional attribute space for events; nodes hold histograms summarizing statistics about the value of this attribute. Assuming sensors lie in a two-dimensional spatial domain, DIFS proposes to use a multi-rooted quad-tree to partition the spatial domain. A normal quad-tree recursively decomposes the sensor domain into four equal quadrants. In a multi-rooted quad-tree, each child can have $l = 2, 4, 8, \ldots$ parents. This is designed to alleviate the hot spot situation higher up in the tree. To couple the decomposition of the spatial domain with the indexing of data attributes, DIFS obeys the following rule: the wider the spatial extent an index node knows about, the more constrained is the value range its histogram covers. For example, assume $l = 4$, that is, each node has four parents. Each internal index node V stores a histogram of value counts in all of its children. Each of V's four parents covers four times the area but indexes only one-fourth of the attribute value range of V. Each leaf index node points directly to storage nodes

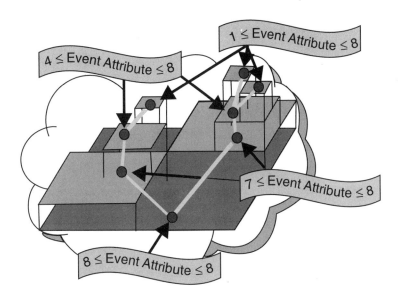

Figure 6.11 An illustration of the DIFS hierarchy (adapted from [83]).

and holds the full range of the attribute. An example is shown in Figure 6.11.

DIFS uses a modified GHT to find an index node. Given a data event and a bounding box, a geographically bounded hash returns a location within that box. This can be done by concatenating the higher-order bits of the bounding box with the hash of the data event. To insert an event into the index tree, the hash function is called to find a closest leaf index node that covers the value of the event. The index node updates its histogram and then forwards the event to its parent, and so on. To query for events in a value range, one first determines the minimum number of index nodes that exactly cover the query range, and then routes the query components to the corresponding index nodes, in order to retrieve the values contained in the subtrees rooted at these nodes. This search for the covering nodes is conducted over the entire spatial domain and can be expensive. DIFS uses the multi-rooted index tree to spread the index data structure across the network in order to load-balance traffic and reduce energy depletion across the network.

6.8.3 Fractional Cascading

The data access patterns of queries in a sensor network are not arbitrary. In many contexts, a query injected in a node of the network is much more likely to request data obtained in the spatial and temporal locality of the query, rather than some other arbitrary data. This was certainly the case when we discussed collaboration groups for target tracking in Chapter 5. More generally, this is because sensor networks measure physical phenomena and physical interactions are mostly localized in space and/or time. In fact, one of the criticisms of the GHT approach is that it may store data away from the node that generated it, far from the vicinity in which that data is most likely to be requested.

This idea of storing information locally is at the heart of the *fractional cascading* approach to storing data in a sensor network [75]. This scheme is also fully symmetric—at the cost of some information duplication, no nodes play a special role any longer, thus avoiding the hot spot issues discussed earlier. The key idea of the fractional cascading approach is to store at each sensor information about data available elsewhere in the network, but in such a way that a sensor knows only a "fraction" of the information from distant parts of the network, in an exponentially decaying fashion by distance. The precise way in which information is to be subsampled, compressed, or aggregated to meet this constraint will be application-dependent. This accomplishes three goals simultaneously:

- The total amount of information duplication across all sensors is kept small, because of the geometric decrease with distance.

- The communication costs required to build this index and its update cost remain reasonable, as on the average information travels only short distances.

- Neighboring sensors have highly correlated world views; this allows for smooth information gradients and enables local search algorithms to work well.

The name *fractional cascading* is aimed to convey this geometric decay of information with distance. The technique automatically adapts the resolution at which information is stored so that more detailed information is available about data obtained in the spatio-temporal locality of the sensor where the query is injected—but without sacrificing the ability to query distant regions or times as well. Furthermore, this is accomplished with load balancing across the entire network field.

To illustrate the potential of fractional cascading, consider the simple scenario of a dense uniform sensor network sampling a smooth scalar field, such as temperature. A typical range query might be to give a query region (specified as a rectangle parallel with the axes) plus a threshold temperature and then ask for all the hot spots—that is, the sensors in the query region where the temperature is above this threshold. For simplicity, we deal here with the static field case only.

We start again with a standard hierarchical structure, say a quad-tree over the sensor field. The root node is associated with a bounding square that covers all the sensors, and we assume that leaf nodes contain at most one sensor each. We associate with each node u in the quad-tree the maximum temperature sensed by any sensor within its bounding square—these are the canonical subsets of the sensor nodes for this method. The fractional cascading storage scheme then stores this fact in all sensor nodes lying within the three siblings of u in the quad-tree. All this can be accomplished by a simple bottom-up process, in which each quad-tree node u is represented by a sensor node within its bounding square. The end result of this precomputation step is that each sensor node stores the maximum temperature for the siblings of each of the ancestor nodes of its corresponding leaf in the quad-tree. A node's view of the world after this propagation is shown in Figure 6.12. Note that the sensor field has been aggregated into larger and larger areas as we move away from the node.

It is not hard to see that with this scheme, in a field of n nodes, each node stores $O(\log n)$ information and the total cost of computing this structure is $O(n \log n)$. In order to process the range query mentioned earlier (finding hot spots in a region R), the node where

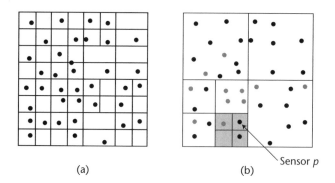

Figure 6.12 (a) Leaves of the quad-tree; (b) a sensor p's view of the world.

the query is injected (source node) proceeds as follows. The source generates a packet to be routed toward the query region. As the query approaches the region R, more detailed information about sensor readings in R becomes available and portions of R may be safely pruned from further consideration (because they are known to be "cold"). In the worst case, our query may need to generate subqueries to every canonical piece inside R and, for those containing hot nodes, traverse a tree to reach all hot spots in that piece. The worst-case cost of the query can be shown to be $O(D + \sqrt{Ak} + P \log P)$, where k is the number of hot sensors in the answer, D is the shortest path from the source to the query range R, and P and A are, respectively, the perimeter and area of R. In a certain model of communication and computation, this can be proven to be nearly best possible.

Note that this improves upon restricted geographic flooding (send the query to everyone in R—that would cost $O(D + A)$), yet still the communication cost of visiting the hot spots and canonical pieces dominates the total query cost. Note also that the method will be especially fast when the query region R is local and small.

6.8.4 Locality-Preserving Hashing

Restricted geographic flooding is possible when the range of interest is specified geometrically. But what if we wanted to do something analogous for regions in some other, arbitrary attribute space?

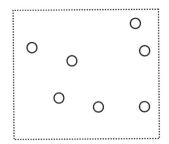 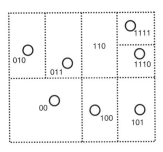 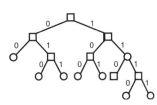

Figure 6.13   A zone tree partitions a space into zones: (a) sensor nodes; (b) zones; (c) corresponding zone tree (adapted from [137]).

Can we have a way to map that (possibly high-dimensional) attribute space to the plane so that nearby locations in attribute space correspond to nearby locations in the plane? This is exactly what *locality-preserving hash functions* [139, 103] accomplish, at least over certain scales. It is these locality-preserving hash functions that enable GHT-like ideas to be used with range queries. This approach is developed in the DIM system [137] (DIM stands for "distributed index for multidimensional data").

DIM generalizes DIFS to index multidimensional data. The key idea of DIM is a clever construction of a locality-preserving mapping between the multidimensional attribute space and the spatial domain of sensors. More specifically, the spatial domain is divided into zones by a recursive spatial bisection alternating between the x and y axes of the spatial domain, as in a k-d tree. Figure 6.13 shows an example of the zone partition. The corresponding zone tree of a space encodes the partition. The path from the root to a zone uniquely identifies the zone with a bit string, the so-called *zone code*, in which 0 represents the left child and 1 the right child at each split in the tree.

The mapping between the attribute space and the spatial domain is accomplished by a locality-preserving geographic hash, in which data with values close to one another are hashed to locations nearby. The hash works in a manner analogous to a k-d tree. Assume each attribute of the data is normalized to between 0 and 1, and the level ℓ of the spatial partition tree is an integer multiple of the dimension d of the data space. The hash function assigns a zone code to a multi-attribute datum as follows: for the first d zone bits, the ith bit is set to 0 if the ith attribute is in [0,0.5]; otherwise it is 1. This process then repeats on the next d bits of the partition by looking at the second bit of the attribute value, meaning if the attribute is in [0, 0.25] or [0.5, 0.75], then the bit is 0; otherwise it is 1. This continues until all the zone code bits are specified.[2] Essentially, the hashing uses the values of the attribute in a round-robin fashion on the zone tree to generate a zone code for the data, similar to a k-d tree.

Inserting data amounts to computing the hash function to determine the corresponding zone and then routing the data to the zone owner, (usually) the node inside the zone. When answering a range query, DIM first finds a node in the zone tree that covers the entire range query. It routes the query to that node using the GPSR routing protocol. Upon receiving a query, a node S examines it to see if the attribute space P it indexes intersects the range specified in the query. If not, the query is sent along to the destination. If so, the node S recursively splits the query range on its subtree, until the subqueries are either completely inside P or completely outside P. Drilling down the subtree rooted at S corresponds to partitioning the query into subqueries using the split-value at each node. The subqueries that are entirely contained in P can be resolved by S locally, and the data is reported back to the query source. The subqueries that are outside P are routed to their respective zones, computed using the hash described earlier. This mirrors the range search in a k-d tree we discussed earlier, except that DIM proposes the use of an initial routing toward the part of the tree that contains the range when the source query is far away from the data; this is to avoid unnecessary communication.

6.9 Temporal Data

Unlike traditional databases that largely store static information, sensor network databases must deal with continuous data acquisition

---

2 The hashing can be generalized to zone trees that are not balanced (see [137]).

and allow the temporal aspects of the data to be used in queries. Temporal databases have been investigated by the database community [216], but in the resource-constrained sensor network context several additional issues arise. These again can be at both the physical and the logical levels. The good news is that sensor data acquired by a single node over time can be processed and summarized directly on that node, with no additional communication costs. The bad news, however, is that overall node storage is very limited. Thus older information must be compressed more and more and eventually discarded, to make room for new data. At the logical level, we may want to allow the user to make queries about the past, the present, or the future (using some predictive mechanism). Even if we are only interested in the present, a predictive mechanism can be useful, since queries can experience long network delays and therefore the data reported is likely to be somewhat stale, anyway. Unlike other temporal databases (like those dealing with stock market data, say), sensor network databases can aim to exploit the continuity of physical phenomena over time in designing effective prediction mechanisms.

6.9.1 Data Aging

The DIMENSIONS system presented in Section 6.8.1 computes multiresolution summaries of data. We discussed multiresolution summarization over the spatial domain, but the same can be done (within each sensor node separately) over the temporal domain. These temporal summaries can then also be aggregated over the spatial domain. As time passes and storage space is needed for new data, older data can be discarded and only some of the temporal summaries maintained. This allows queries about the past to access aggregated useful information.

The exact way this happens is application-dependent. In DIMENSIONS, the aging strategy is defined via a (user-specified) *aging function*, a monotonically decreasing function specifying acceptable query response accuracy (or, equivalently, summary fidelity) as a function of data age. Given the hierarchical structure of data aggregation, a schedule for discarding data and data summaries must be computed. This reduces to solving a constrained optimization problem, expressing the need to meet the device storage constraints and best satisfy the aging function.

The design of good aging functions is a nontrivial task that can benefit from data mining over past data and queries, or from running an optimization procedure over a training data set. In the absence of prior data, a greedy algorithm can be used that assigns weights to summaries according to their expected importance during drill-down queries.

6.9.2 Indexing Motion Data

Because so much of what sensor networks capture is continuously evolving phenomena, an important research area is how to build indices for the corresponding continuously changing sensor data. A fixed index structure will soon be obsolete, while continuous updates to a static index can incur heavy modification and communication costs. Furthermore, it is highly desirable to allow temporal queries directly in the system. For example, if we are tracking a set of vehicles moving over a terrain, we may want to ask for all vehicles inside a region R at time t, for various values of R and t, or for an interval of time Δt. Even predictive queries about the future are desirable, based on the best available information we have about the vehicles and their motion plans. Naturally, the more distant this future is, the greater the inaccuracy we must allow in the query answers. Languages for expressing temporal queries have been investigated in [185, 89].

There are two general approaches to indexing motion data. In the case where accurate motion information about the objects is known in advance, a time-oblivious approach can be taken. In this approach, time is simply viewed as another attribute dimension and the objects being indexed are the known object trajectories. In our previous vehicle example, we have a three-dimensional attribute space with two spatial and one temporal dimension; each vehicle corresponds to a curve in this space. Geometric methods can then be used directly on that space [185, 181] or, if the vehicle trajectories can be parameterized with a small number of parameters (e.g., linear motions), in a derived configuration space where each trajectory becomes a point [118, 6]. An index for such a configuration space allows queries about the past, present, or future equally well. It needs to be modified only when new objects are inserted or deleted, or when the trajectory of an object changes. Both the index construction and updates can be quite expensive, however, because the dimensionality of the space may be high.

In the physical world, it is unlikely that we will have full prior knowledge of the temporal evolution of the phenomenon of interest—this can be exactly what the sensor network is there to capture. An alternative approach is to maintain a dynamic index on the current state of the world, but one that needs to be updated only when certain critical events occur. This is the approach taken by the Kinetic Data Structures (KDS) framework [86, 16]. In the KDS setting, the correctness of the index is certified by certain atomic predicates, called *certificates*, on the parameters defining the index. For instance, in a range-searching setting, if we want to build a *k-d* tree on the vehicle positions in the plane, a certificate might be that the *x*-coordinate of vehicle *v* is less than the partitioning value *s*. Altogether, a collection of such certificates—the KDS *assertion cache*— guarantees the validity of the *k-d* tree, which can then be used to answer range queries on the vehicle positions even as the vehicles move, as long as these certificates remain true. When a sensor detects that a certificate has failed, a KDS repair mechanism is invoked that will update the *k-d* tree if necessary, as well as its associated assertion cache [7]. If the certificate assertions are chosen carefully, we can hope that this repair procedure will be a local, incremental update to the index. Effectively, a KDS attempts to translate the continuity or coherence of motion in the world to such localized updates to the index. This certificate or relation-based view of index updates is closely related to relational tracking, as discussed in Section 5.3.4.

The KDS approach incrementally tracks the index structure as objects move and can be used to answer queries about the current state of the world. An easy extension allows for queries arriving in chronological order, that is, queries whose time stamps are in nondecreasing order. Using more complicated persistence techniques [59], the index can also answer queries about states of the world in the near past or the near future. Because the KDS approach requires that the index be updated as time goes on in order to exploit coherence, it may lead to wasted processing during periods of inactivity, when no queries are present in the network. Furthermore, exactly how to apportion the certificate state across nodes and how to perform the cache update in a distributed fashion are problems that have not yet received the needed research attention.

The two approaches can be combined to get the benefits of both. In many applications, such as traffic monitoring, it is most crucial to answer queries related to near future configurations of the environment. The idea is to add time as an additional dimension but optimize the index for times close to the present, while storing only an approximate representation of the world state in the more distant future. The index is then updated periodically using a KDS approach to maintain this invariant; these updates need not be so frequent if the motion predictions are accurate. Such time-responsive indexing has been used in a number of recent database papers [6, 199, 8]. For further details, the reader is referred to the excellent survey [5].

6.10 Summary

This chapter introduced the general area of sensor network databases and discussed the fundamental issues and trade-offs in designing such systems. Along the way, we presented techniques for SQL-style query interfaces, in-network aggregation, range searching, distributed index construction, and handling temporal data. A key take-home message is that energy optimization, robustness to failures, scalability, and load balancing are key considerations in the design. As the survey of the literature also makes clear, this area is still in its infancy, and much more needs to be done to bring sensor network databases to a mature level.

The basic aggregation functions we presented may be extended to support more sophisticated data analysis. What are appropriate summaries of sensor network data remains a topic for further research. Geometric computation is an inherent part of data analysis in many sensor network applications such as topographic mapping. For example, to compute iso-contours of an elevation map, one needs to introduce a few geometric primitives on a grid of data, similar to the marching cubes method in computer graphics. More generally, finite-element-style computation may be supported on a triangulation of the sensor grid.

Another important area of further research concerns query handoff and subquery initiation. In a vehicle-tracking application, we may desire each node to start tracking only when a vehicle is nearby and stop the query when the vehicle moves away, due to energy considerations. This requires each node to be able to hand off its query to another node, a problem we studied in Chapter 5. In other cases, one may desire to split the current query into multiple subqueries, one for each subtask—we already saw this during drill down, in most of the hierarchical indices we discussed. As another example, one may conclude that a vehicle is likely to have taken two possible courses, one going straight through an intersection of roads and one making a right turn. To disambiguate between the two possibilities, the current vehicle-tracking query can be split into two subqueries, one for each possible direction, and the answer is synthesized after the subqueries return. Future research should address how query processing can leverage sensor tasking and signal processing, and specify appropriate APIs between these modules. Some of these issues are discussed in the context of TinyDB in [149].

Finally, as we remarked, integration of query processing with the networking layer, the mapping of index structures to the spatial topology of the network, and distributed index construction for motion data all remain important topics for further investigation.

7

Sensor Network Platforms and Tools

In previous chapters, we discussed various aspects of sensor networks, including sensing and estimation, networking, infrastructure services, sensor tasking, and data storage and query. A real-world sensor network application most likely has to incorporate all these elements, subject to energy, bandwidth, computation, storage, and real-time constraints. This makes sensor network application development quite different from traditional distributed system development or database programming. With ad hoc deployment and frequently changing network topology, a sensor network application can hardly assume an always-on infrastructure that provides reliable services such as optimal routing, global directories, or service discovery.

There are two types of programming for sensor networks, those carried out by end users and those performed by application developers. An end user may view a sensor network as a pool of data and *interact* with the network via queries. Just as with query languages for database systems like SQL, a good sensor network programming language should be expressive enough to encode application logic at a high level of abstraction, and at the same time be structured enough to allow efficient execution on the distributed platform. Examples of sensor database query interfaces are described in Chapter 6. Ideally, the end users should be shielded away from details of how sensors are organized and how nodes communicate.

On the other hand, an application developer must provide end users of a sensor network with the capabilities of data acquisition, processing, and storage. Unlike general distributed or database systems, collaborative signal and information processing (CSIP) software comprises reactive, concurrent, distributed programs running on ad hoc, resource-constrained, unreliable computation and communication platforms. Developers at this level have to deal with all kinds of uncertainty in the real world. For example, signals are noisy, events can happen at the same time, communication and computation take time, communications may be unreliable, battery life is limited, and so on. Moreover, because of the amount of domain knowledge required, application developers are typically signal and information processing specialists, rather than operating systems and networking experts. How to provide appropriate programming abstractions to these application writers is a key challenge for sensor network software development. In this chapter, we focus on software design issues to support this type of programming.

To make our discussion of these software issues concrete, we first give an overview of a few representative sensor node hardware platforms (Section 7.1). In Section 7.2, we present the challenges of sensor network programming due to the massively concurrent interaction with the physical world. Section 7.3 describes TinyOS for Berkeley motes and two types of node-centric programming interfaces: an imperative language, nesC, and a dataflow-style language, TinyGALS. Node-centric designs are typically supported by node-level simulators such as ns-2 and TOSSIM, as described in Section 7.4. State-centric programming is a step toward programming beyond individual nodes. It gives programmers platform support for thinking in high-level abstractions, such as the state of the phenomena of interest over space and time. An example of state-centric platforms is given in Section 7.5.

7.1 Sensor Node Hardware

Sensor node hardware can be grouped into three categories, each of which entails a different set of trade-offs in the design choices.

- *Augmented general-purpose computers:* Examples include low-power PCs, embedded PCs (e.g., PC104), custom-designed PCs (e.g., Sensoria WINS NG nodes),[1] and various personal digital assistants (PDA). These nodes typically run off-the-shelf operating systems such as Win CE, Linux, or real-time operating systems and use standard wireless communication protocols such as Bluetooth or IEEE 802.11. Because of their relatively higher processing capability, they can accommodate a wide variety of sensors, ranging from simple microphones to more sophisticated video cameras.

Compared with dedicated sensor nodes, PC-like platforms are more power hungry. However, when power is not an issue, these platforms have the advantage that they can leverage the availability of fully supported networking protocols, popular programming languages, middleware, and other off-the-shelf software.

- *Dedicated embedded sensor nodes:* Examples include the Berkeley mote family [98], the UCLA Medusa family [202], Ember nodes,[2] and MIT μAMP [32]. These platforms typically use commercial off-the-shelf (COTS) chip sets with emphasis on small form factor, low power processing and communication, and simple sensor interfaces. Because of their COTS CPU, these platforms typically support at least one programming language, such as C. However, in order to keep the program footprint small to accommodate their small memory size, programmers of these platforms are given full access to hardware but barely any operating system support. A classical example is the TinyOS platform and its companion programming language, nesC. We will discuss these platforms in Sections 7.3.1 and 7.3.2.

- *System-on-chip (SoC) nodes:* Examples of SoC hardware include smart dust [109], the BWRC picoradio node [187], and the PASTA node.[3] Designers of these platforms try to push the hardware limits by fundamentally rethinking the hardware architecture trade-offs for a sensor node at the chip design level. The goal is to find new ways of integrating CMOS, MEMS, and RF technologies

---

1 See *http://www.sensoria.com/* and *http://www.janet.ucla.edu/WINS/*, and [158].
2 See *http://www.ember.com*.
3 See *http://pasta.east.isi.edu*.

to build extremely low power and small footprint sensor nodes that still provide certain sensing, computation, and communication capabilities. Since most of these platforms are currently in the research pipeline with no predefined instruction set, there is no software platform support available.

Among these hardware platforms, the Berkeley motes, due to their small form factor, open source software development, and commercial availability, have gained wide popularity in the sensor network research community. In the following section, we give an overview of the Berkeley MICA mote.

7.1.1 Berkeley Motes

The Berkeley motes are a family of embedded sensor nodes sharing roughly the same architecture. Figure 7.1 shows a comparison of a subset of mote types.

Mote type		WeC	Rene	Rene2	Mica	Mica2	Mica2Dot
Example picture							
MCU	Chip	AT90LS8535	ATmega163L		ATmega103L	ATmega128L	
	Type	4 MHz, 8 bit	4 MHz, 8 bit		4 MHz, 8 bit	8 MHz, 8 bit	
	Program memory (KB)	8	16		128	128	
	RAM (KB)	0.5	1		4	4	
External nonvolatile storage	Chip	24LC256			AT45DB014B		
	Connection type	I2C			SPI		
	Size (KB)	32			512		
Default power source	Type	Coin cell			2xAA		Coin cell
	Typical capacity (mAh)	575			2850		1000
RF	Chip	TR1000				CC1000	
	Radio frequency	868/916MHz				868/916MHz, 433, or 315 MHz	
	Raw speed (kbps)	10			40	38.4	
	Modulation type	On/Off key			Amplitude Shift key	Frequency Shift key	

Figure 7.1 A comparison of Berkeley motes.

Let us take the MICA mote as an example. The MICA motes have a two-CPU design, as shown in Figure 7.2. The main microcontroller (MCU), an Atmel ATmega103L, takes care of regular processing. A separate and much less capable coprocessor is only active when the MCU is being reprogrammed. The ATmega103L MCU has integrated 512 KB flash memory and 4 KB of data memory. Given these small memory sizes, writing software for motes is challenging. Ideally, programmers should be relieved from optimizing code at assembly level to keep code footprint small. However, high-level support and software services are not free. Being able to mix and match only necessary software components to support a particular application is essential to achieving a small footprint. A detailed discussion of the software architecture for motes is given in Section 7.3.1.

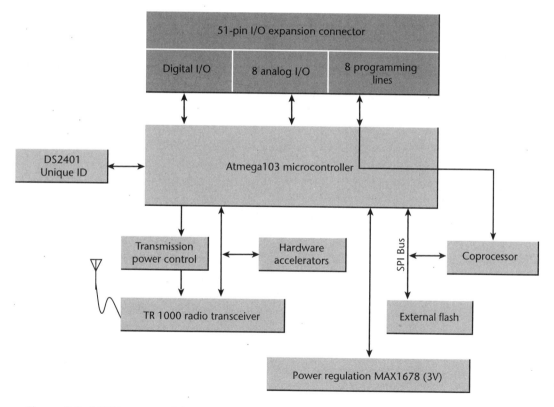

Figure 7.2 MICA mote architecture.

In addition to the memory inside the MCU, a MICA mote also has a separate 512 KB flash memory unit that can hold data. Since the connection between the MCU and this external memory is via a low-speed serial peripheral interface (SPI) protocol, the external memory is more suited for storing data for later batch processing than for storing programs. The RF communication on MICA motes uses the TR1000 chip set (from RF Monolithics, Inc.) operating at 916 MHz band. With hardware accelerators, it can achieve a maximum of 50 kbps raw data rate. MICA motes implement a 40 kbps transmission rate. The transmission power can be digitally adjusted by software though a potentiometer (Maxim DS1804). The maximum transmission range is about 300 feet in open space.

Like other types of motes in the family, MICA motes support a 51 pin I/O extension connector. Sensors, actuators, serial I/O boards, or parallel I/O boards can be connected via the connector. A sensor/actuator board can host a temperature sensor, a light sensor, an accelerometer, a magnetometer, a microphone, and a beeper. The serial I/O (UART) connection allows the mote to communicate with a PC in real time. The parallel connection is primarily for downloading programs to the mote.

It is interesting to look at the energy consumption of various components on a MICA mote. As shown in Figure 7.3, a radio

Component	Rate	Startup time	Current consumption
MCU active	4 MHz	N/A	5.5 mA
MCU idle	4 MHz	1 µs	1.6 mA
MCU suspend	32 kHz	4 ms	<20 µA
Radio transmit	40 kHz	30 ms	12 mA
Radio receive	40 kHz	30 ms	1.8 mA
Photoresister	2000 Hz	10 ms	1.235 mA
Accelerometer	100 Hz	10 ms	5 mA/axis
Temperature	2 Hz	500 ms	0.150 mA

Figure 7.3   Power consumption of MICA motes.

transmission bears the maximum power consumption. However, each radio packet (e.g., 30 bytes) only takes 4 ms to send, while listening to incoming packets turns the radio receiver on all the time. The energy that can send one packet only supports the radio receiver for about 27 ms. Another observation is that there are huge differences among the power consumption levels in the active mode, the idle mode, and the suspend mode of the MCU. It is thus worthwhile from an energy-saving point of view to suspend the MCU and the RF receiver as long as possible.

7.2 Sensor Network Programming Challenges

Traditional programming technologies rely on operating systems to provide abstraction for processing, I/O, networking, and user interaction hardware, as illustrated in Figure 7.4. When applying such a

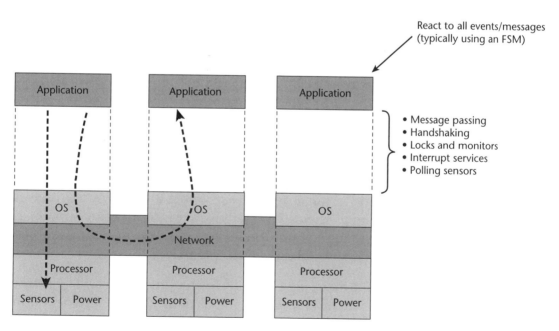

Figure 7.4 Traditional embedded system programming interface.

model to programming networked embedded systems, such as sensor networks, the application programmers need to explicitly deal with message passing, event synchronization, interrupt handing, and sensor reading. As a result, an application is typically implemented as a finite state machine (FSM) that covers all extreme cases: unreliable communication channels, long delays, irregular arrival of messages, simultaneous events, and so on. In a target tracking application implemented on a Linux operating system and with directed diffusion routing, roughly 40 percent of the code implements the FSM and the glue logic of interfacing computation and communication [142].

For resource-constrained embedded systems with real-time requirements, several mechanisms are used in embedded operating systems to reduce code size, improve response time, and reduce energy consumption. Microkernel technologies [211] modularize the operating system so that only the necessary parts are deployed with the application. Real-time scheduling [27] allocates resources to more urgent tasks so that they can be finished early. Event-driven execution allows the system to fall into low-power sleep mode when no interesting events need to be processed. At the extreme, embedded operating systems tend to expose more hardware controls to the programmers, who now have to directly face device drivers and scheduling algorithms, and optimize code at the assembly level. Although these techniques may work well for small, stand-alone embedded systems, they do not scale up for the programming of sensor networks for two reasons.

- Sensor networks are large-scale distributed systems, where global properties are derivable from program execution in a massive number of distributed nodes. Distributed algorithms themselves are hard to implement, especially when infrastructure support is limited due to the ad hoc formation of the system and constrained power, memory, and bandwidth resources.

- As sensor nodes deeply embed into the physical world, a sensor network should be able to respond to multiple concurrent stimuli at the speed of changes of the physical phenomena of interest.

In the rest of the chapter, we give several examples of sensor network software design platforms. We discuss them in terms of both *design methodologies* and *design platforms*. A design methodology implies a conceptual model for programmers, with associated techniques for problem decomposition for the software designers. For example, does the programmer think in terms of events, message passing, and synchronization, or does he/she focus more on information architecture and data semantics? A design platform supports a design methodology by providing design-time (precompile time) language constructs and restrictions, and run-time (postcompile time) execution services.

There is no single universal design methodology for all applications. Depending on the specific tasks of a sensor network and the way the sensor nodes are organized, certain methodologies and platforms may be better choices than others. For example, if the network is used for monitoring a small set of phenomena and the sensor nodes are organized in a simple star topology, then a client-server software model would be sufficient. If the network is used for monitoring a large area from a single access point (i.e., the base station), and if user queries can be decoupled into aggregations of sensor readings from a subset of sensor nodes, then a tree structure that is rooted at the base station is a better choice. However, if the phenomena to be monitored are moving targets, as in the target tracking examples discussed in Chapter 2, then neither the simple client-server model nor the tree organization is optimal. More sophisticated design methodologies and platforms are required.

7.3 Node-Level Software Platforms

Most design methodologies for sensor network software are node-centric, where programmers think in terms of how a node should behave in the environment. A node-level platform can be a node-centric operating system, which provides hardware and networking abstractions of a sensor node to programmers, or it can be a language platform, which provides a library of components to programmers.

A typical operating system abstracts the hardware platform by providing a set of services for applications, including file management, memory allocation, task scheduling, peripheral device drivers, and networking. For embedded systems, due to their highly specialized applications and limited resources, their operating systems make different trade-offs when providing these services. For example, if there is no file management requirement, then a file system is obviously not needed. If there is no dynamic memory allocation, then memory management can be simplified. If prioritization among tasks is critical, then a more elaborate priority scheduling mechanism may be added.

TinyOS [98] and TinyGALS [38] are two representative examples of node-level programming tools that we will cover in detail in this section. Other related software platforms include Maté [130], a virtual machine for the Berkeley motes. Observing that operations such as polling sensors and accessing internal states are common to all sensor network application, Maté defines virtual machine instructions to abstract those operations. When a new hardware platform is introduced with support for the virtual machine, software written in the Maté instruction set does not have to be rewritten.

7.3.1 Operating System: TinyOS

TinyOS aims at supporting sensor network applications on resource-constrained hardware platforms, such as the Berkeley motes.

To ensure that an application code has an extremely small footprint, TinyOS chooses to have no file system, supports only static memory allocation, implements a simple task model, and provides minimal device and networking abstractions. Furthermore, it takes a language-based application development approach, to be discussed later, so that only the necessary parts of the operating system are compiled with the application. To a certain extent, each TinyOS application is built into the operating system.

Like many operating systems, TinyOS organizes components into layers. Intuitively, the lower a layer is, the "closer" it is to the hardware; the higher a layer is, the "closer" it is to the application.

In addition to the layers, TinyOS has a unique component architecture and provides as a library a set of system software components. A component specification is independent of the component implementation. Although most components encapsulate software functionalities, some are just thin wrappers around hardware. An application, typically developed in the nesC language covered in the next section, *wires* these components together with other application-specific components.

Let us consider a TinyOS application example—FieldMonitor, where all nodes in a sensor field periodically send their temperature and photo sensor readings to a base station via an ad hoc routing mechanism. A diagram of the FieldMonitor application is shown in Figure 7.5, where blocks represent TinyOS components and arrows represent function calls among them. The directions of the arrows are from callers to callees.

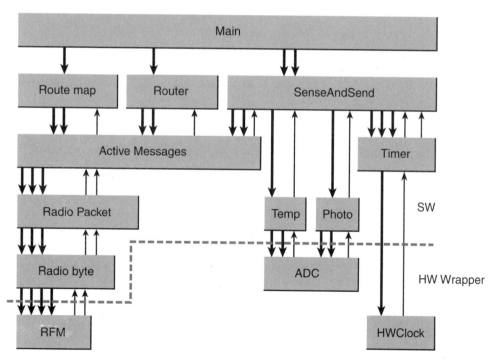

Figure 7.5 The FieldMonitor application for sensing and sending measurements.

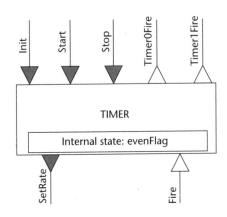

Figure 7.6   The Timer component and its interfaces.

To explain in detail the semantics of TinyOS components, let us first look at the Timer component of the FieldMonitor application, as shown in Figure 7.6. This component is designed to work with a clock, which is a software wrapper around a hardware clock that generates periodic interrupts. The method calls of the Timer component are shown in the figure as the arrowheads. An arrowhead pointing into the component is a method of the component that other components can call. An arrowhead pointing outward is a method that this component requires another layer component to provide. The absolute directions of the arrows, up or down, illustrate this component's relationship with other layers. For example, the Timer depends on a lower layer HWClock component. The Timer can set the rate of the clock, and in response to each clock interrupt it toggles an internal Boolean flag, evenFlag, between true (or 1) and false (or 0). If the flag is 0, the Timer produces a timer0Fire event to trigger other components; otherwise, it produces a timer1Fire event. The Timer has an init() method that initializes its internal flag, and it can be enabled and disabled via the start and stop calls.

A program executed in TinyOS has two contexts, *tasks* and *events*, which provide two sources of concurrency. Tasks are created (also called *posted*) by components to a task scheduler. The default implementation of the TinyOS scheduler maintains a task queue and invokes tasks according to the order in which they were posted. Thus tasks are deferred computation mechanisms. Tasks always run to completion without preempting or being preempted by other tasks. Thus tasks are nonpreemptive. The scheduler invokes a new task from the task queue only when the current task has completed. When no tasks are available in the task queue, the scheduler puts the CPU into the sleep mode to save energy.

The ultimate sources of triggered execution are events from hardware: clock, digital inputs, or other kinds of interrupts. The execution of an interrupt handler is called an *event context*. The processing of events also runs to completion, but it preempts tasks and can be preempted by other events. Because there is no preemption mechanism among tasks and because events always preempt tasks, programmers are required to chop their code, especially the code in the event contexts, into small execution pieces, so that it will not block other tasks for too long.

Another trade-off between nonpreemptive task execution and program reactiveness is the design of split-phase operations in TinyOS. Similar to the notion of asynchronous method calls in distributed computing, a split-phase operation separates the initiation of a method call from the return of the call. A call to a split-phase operation returns immediately, without actually performing the body of the operation. The true execution of the operation is scheduled later; when the execution of the body finishes, the operation notifies the original caller through a separate method call. An example of a split-phase operation is the packet send method in the Active Messages (AM) component, used in Figure 7.5. Sending a packet is a long operation, involving converting the packets to bytes, then to bits, and ultimately driving the RF circuits to send the bits one by one. Without a split-phase execution, sending a packet will block the entire system from reacting to new events for a significant period of time. In the TinyOS implementation, the send() command in the AM component returns immediately. However, it is the caller's responsibility to remember that the packet has not yet been sent. When the packet is indeed sent, the AM component will notify its caller by a sendDone() method call. Only at this time is the AM component ready to accept another packet.

In TinyOS, resource contention is typically handled through explicit rejection of concurrent requests. All split-phase operations return Boolean values indicating whether a request to perform the operation is accepted. In the above example, a call of send(), when the AM component is still sending the first packet, will result in an error signaled by the AM component. To avoid such an error, the caller of the AM component typically implements a *pending* lock, to remember not to request further sendings until the sendDone() method is called. To avoid loss of packets, a queue should be incorporated by the caller if necessary.

In summary, many design decisions in TinyOS are made to ensure that it is extremely lightweight. Using a component architecture that contains all variables inside the components and disallowing dynamic memory allocation reduces the memory management overhead and makes the data memory usage statically analyzable. The simple concurrency model allows high concurrency with low thread maintenance overhead. As a consequence, the entire FieldMonitor system shown in Figure 7.5 takes only 3 KB of space for code and 226 bytes for data. However, the advantage of being lightweight is not without cost. Many hardware idiosyncrasies and complexities of concurrency management are left for the application programmers to handle. Several tools have been developed to give programmers language-level support for improving programming productivity and code robustness. We introduce in the next two sections two special-purpose languages for programming sensor network nodes. Although both languages are designed on top of TinyOS, the principles they represent may apply to other platforms.

7.3.2 Imperative Language: nesC

nesC [79] is an extension of C to support and reflect the design of TinyOS v1.0 and above. It provides a set of language constructs and restrictions to implement TinyOS components and applications.

Component Interface

A component in nesC has an interface specification and an implementation. To reflect the layered structure of TinyOS, interfaces of a nesC component are classified as *provides* or *uses* interfaces. A provides interface is a set of method calls exposed to the upper layers, while a uses interface is a set of method calls hiding the lower layer components. Methods in the interfaces can be grouped and named. For example, the interface specification of the Timer component in Figure 7.6 is listed in Figure 7.7. The interface, again, independent of the implementation, is called TimerModule.

Although they have the same method call semantics, nesC distinguishes the *directions* of the interface calls between layers as *event* calls

```
module TimerModule {
  provides {
    interface StdControl;
    interface Timer01;
  }
  uses interface Clock as Clk;
}

interface StdControl {
  command result_t init();
}

interface Timer01 {
  command result_t start(char type, uint32_t interval;
  command result_t stop();
  event result_t timer0Fire();
  event result_t timer1Fire();
}

interface Clock {
  command result_t setRate(char interval, char scale);
  event result_t fire();
}
```

Figure 7.7 The interface definition of the Timer component in nesC.

and *command* calls. An event call is a method call from a lower layer component to a higher layer component, while a command is the opposite. Note that one needs to know both the type of the interface (provides or uses) and the direction of the method call (event or command) to know exactly whether an interface method is implemented by the component or is required by the component.

The separation of interface type definitions from how they are used in the components promotes the reusability of standard interfaces. A component can provide and use the same interface type, so that it can act as a filter interposed between a client and a service. A component may even use or provide the same interface multiple times. In these cases, the component must give each interface instance a separate name using the as notation, as shown in the Clock interface in Figure 7.7.

Component Implementation

There are two types of components in nesC, depending on how they are implemented: *modules* and *configurations*. Modules are implemented by application code (written in a C-like syntax). Configurations are implemented by connecting interfaces of existing components.

The implementation part of a module is written in C-like code. A command or an event bar in an interface foo is referred as foo.bar. A keyword call indicates the invocation of a command. A keyword signal indicates the triggering by an event. For example, Figure 7.8 shows part of the implementation of the Timer component, whose interface is defined in Figure 7.7. In a sense, this implementation is very much like an object in object-oriented programming without any constructors.

Configuration is another kind of implementation of components, obtained by connecting existing components. Suppose we want to connect the Timer component and a hardware clock wrapper, called HWClock, to provide a timer service, called TimerC. Figure 7.9 shows a conceptual diagram of how the components are connected, and Figure 7.10 shows the corresponding nesC code.

```
module Timer {
  provides {
    interface StdControl;
    interface Timer01;
  }
  uses interface Clock as Clk;
}
implementation {
  bool evenFlag;

  command result_t StdControl.init() {
    evenFlag = 0;
    return call Clk.setRate(128, 4); //4 ticks per second
  }

  event result_t Clk.fire() {
    evenFlag = !evenFlag;
    if (evenFlag) {
      signal Timer01.timer0Fire();
    } else {
      signal Timer01.timer1Fire();
    }
    return SUCCESS;
  }
  ...
}
```

Figure 7.8 The implementation definition of the Timer component in nesC.

First of all, notice that the keyword configuration in the specification indicates that this component is not implemented directly as a module. In the implementation section of the configuration, the code first includes the two components, and then specifies that the interface StdControl of the TimerC component is the StdControl interface of the TimerModule; similarly for the Timer01 interface. The connection between the Clock interfaces is specified using the -> operator. Essentially, this interface is hidden from upper layers.

nesC also supports the creation of several instances of a component by declaring *abstract components* with optional parameters. Abstract components are created at compile time in configurations.

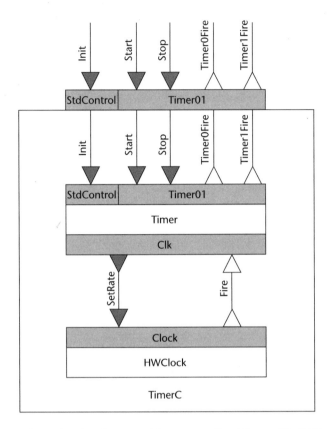

Figure 7.9   The TimerC configuration implemented by connecting Timer with HWClock.

Recall that TinyOS does not support dynamic memory allocation, so all components are statically constructed at compile time.

A complete application is always a configuration rather than a module. An application must contain the Main module, which links the code to the scheduler at run time. The Main has a single StdControl interface, which is the ultimate source of initialization of all components.

Concurrency and Atomicity

The language nesC directly reflects the TinyOS execution model through the notion of command and event contexts. Figure 7.11

```
configuration TimerC {
  provides {
    interface StdControl;
    interface Timer01;
  }
}
implementation {
  components TimerModule, Clock;

  StdControl = TimerModule.StdControl;
  Timer = TimerModule.Timer;

  TimerModule.Clk -> HWClock.Clock;
}
```

Figure 7.10 The implementation definition of the TimerC configuration in nesC.

shows a section of the component SenseAndSend to illustrate some language features to support concurrency in nesC and the effort to reduce race conditions. The SenseAndSend component is intended to be built on top of the Timer component (described in the previous section), an ADC component, which can provide sensor readings, and a communication component, which can send (or, more precisely, broadcast) a packet. When responding to a timer0Fire event, the SenseAndSend component invokes the ADC to poll a sensor reading. Since polling a sensor reading can take a long time, a split-phase operation is implemented for getting sensor readings. The call to ADC.getData() returns immediately, and the completion of the operation is signaled by an ADC.dataReady() event. A busy flag is used to explicitly reject new requests while the ADC is fulfilling an existing request. The ADC.getData() method sets the flag to true, while the ADC.dataReady() method sets it back to false. Sending the sensor reading to the next-hop neighbor via wireless communication is also a long operation. To make sure that it does not block the processing of the ADC.dataReady() event, a separate task is posted to the scheduler. A task is a method defined using the task keyword. In order

```
module SenseAndSend{
  provides interface StdControl;
    uses interface ADC;
    uses interface Timer:
    uses interface Send;
}

implementation {
  bool busy;
  norace uint16_t sensorReading;

  command result_t StdControl.init() {
    busy = FALSE;
  }

  event result_t Timer.timer0Fire() {
    bool localBusy;
    atomic {
      localBusy = busy;
      busy = TRUE;
    }
    if (!localBusy} {
      call ADC.getData(); //start getting sensor reading
      return SUCESS;
    } else {
      return FAILED;
    }
  }

  task void sendData() { // send sensorReading
    adcPacket.data = sensorReading;
    call Send.send(&adcPacket, sizeof adcPacket.data};
    return SUCESS;
  }

  event result_t ADC.dataReady(uinit16_t data) {
    sensorReading = data;
    post sendData();
    atomic {
      busy = FALSE;
    }
    return SUCESS;
  }
  ...
}
```

Figure 7.11 A section of the implementation of SenseAndSend, illustrating the handling of concurrency in nesC.

to simplify the data structures inside the scheduler, a task cannot have arguments. Thus the sensor reading to be sent is put into a sensorReading variable.

There is one source of race condition in the SenseAndSend, which is the updating of the busy flag. To prevent some state from being updated by both scheduled tasks and event-triggered interrupt handlers, nesC provides language facilities to limit the race conditions among these operations.

In nesC, code can be classified into two types:

- *Asynchronous code (AC):* Code that is reachable from at least one interrupt handler.

- *Synchronous code (SC):* Code that is only reachable from tasks.

Because the execution of TinyOS tasks are nonpreemptive and interrupt handlers preempts tasks, SC is always atomic with respect to other SCs. However, any update to shared state from AC, or from SC that is also updated from AC, is a potential race condition. To reinstate atomicity of updating shared state, nesC provides a keyword atomic to indicate that the execution of a block of statements should not be preempted. This construction can be efficiently implemented by turning off hardware interrupts. To prevent blocking the interrupts for too long and affecting the responsiveness of the node, nesC does not allow method calls in atomic blocks. In fact, nesC has a compiler rule to enforce the accessing of shared variables to maintain the race-free condition. If a variable x is accessed by AC, then any access of x outside of an atomic statement is a compile-time error. This rule may be too rigid in reality. When a programmer knows for sure that a data race is not going to occur, or does not care if it occurs, then a norace declaration of the variable can prevent the compiler from checking the race condition on that variable.

Thus, to correctly handle concurrency, nesC programmers need to have a clear idea of what is synchronous code and what is asynchronous code. However, since the semantics is hidden away in the layered structure of TinyOS, it is sometimes not obvious to the programmers where to add atomic blocks.

7.3.3   Dataflow-Style Language: TinyGALS

Dataflow languages [3] are intuitive for expressing computation on interrelated data units by specifying data dependencies among them. A dataflow program has a set of processing units called *actors*. Actors have ports to receive and produce data, and the directional connections among ports are FIFO queues that mediate the flow of data. Actors in dataflow languages intrinsically capture concurrency in a system, and the FIFO queues give a structured way of decoupling their executions. The execution of an actor is triggered when there are enough input data at the input ports.

Asynchronous event-driven execution can be viewed as a special case of dataflow models, where each actor is triggered by every incoming event. The *globally asynchronous and locally synchronous* (GALS) mechanism is a way of building event-triggered concurrent execution from thread-unsafe components. TinyGALS is such a language for TinyOS.

One of the key factors that affects component reusability in embedded software is the component composability, especially concurrent composability. In general, when developing a component, a programmer may not anticipate all possible scenarios in which the component may be used. Implementing all access to variables as atomic blocks incurs too much overhead. At the other extreme, making all variable access unprotected is easy for coding but certainly introduces bugs in concurrent composition. TinyGALS addresses concurrency concerns at the system level, rather than at the component level as in nesC. Reactions to concurrent events are managed by a dataflow-style FIFO queue communication.

TinyGALS Programming Model
TinyGALS supports all TinyOS components, including its interfaces and module implementations.[4] All method calls in a component

---

4 Although posting tasks is not part of the TinyGALS semantics, the TinyGALS compiler and run time are compatible with it.

interface are synchronous method calls—that is, the thread of control enters immediately into the callee component from the caller component. An application in TinyGALS is built in two steps: (1) constructing asynchronous actors from synchronous components,[5] and (2) constructing an application by connecting the asynchronous components though FIFO queues.

An actor in TinyGALS has a set of input ports, a set of output ports, and a set of connected TinyOS components. An actor is constructed by connecting synchronous method calls among TinyOS components. For example, Figure 7.12 shows a construction of TimerActor

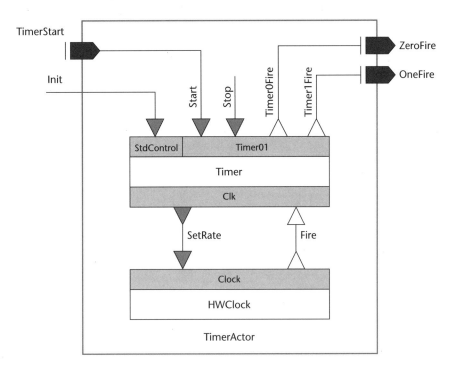

Figure 7.12 Construction of a TimerActor from a Timer component and a Clock component.

---

5 In the implementation of TinyGALS as described in [39], which is based on TinyOS 0.6.1 and predates nesC, the asynchronous actors are called *modules*, and asynchronous connections are represented as "->". To avoid the confusion with nesC, we have modified some of the TinyGALS syntax for inclusion in this section.

from two TinyOS components (i.e., nesC modules), Timer and Clock. Figure 7.13 is the corresponding TinyGALS code. An actor can expose one or more initialization methods. These methods are called by the TinyGALS run time before the start of an application. Initialization methods are called in a nondeterministic order, so their implementations should not have any cross-component dependencies.

At the application level, the asynchronous communication of actors is mediated using FIFO queues. Each connection can be parameterized by a queue size. In the current implementation of TinyGALS, events are discarded when the queue is full. However, other mechanisms such as discarding the oldest event can be used. Figure 7.14 shows a TinyGALS composition of timing, sensing, and sending part of the FieldMonitor application in Figure 7.5.

```
Actor TimerActor {
  include components {
    TimerModule;
    HWClock;
  }
  init {
    TimerModule.init;
  }
  port in {
    timerStart;
  }
  port out {
    zeroFire;
    oneFire;
  }
}
implementation {
  timerStart -> TimerModule.Timer.start;
  TimerModule.Clk -> HWClock.Clock;
  TimerModule.Timer.timer0Fire -> zeroFire;
  TimerModule.Timer.timer1Fire -> oneFire;
}
```

Figure 7.13 Implementation of the TimerActor in TinyGALS.

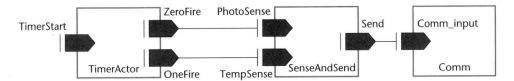

Figure 7.14 Triggering, sensing, and sending actors of the FieldMonitor in TinyGALS.

```
Application FieldMonitor {
  include actors {
    TimerActor;
    SenseAndSend;
    Comm;
  }
  implementation {
    zeroFire => photoSense 5;
    oneFire => tempSense 5;
    send => comm_input 10;
  }
  START@ timerStart;
}
```

Figure 7.15 Implementation of the FieldMonitor in TinyGALS.

Figure 7.15 is the TinyGALS specification of the configuration in Figure 7.14. We omit the details of the SenseAndSend actor and the Comm actor, whose ports are shown in Figure 7.14. The symbol => represents a FIFO queue connecting input ports and output ports. The integer at the end of the line specifies the queue size. The command START@ indicates that the TinyGALS run time puts an initial event into the corresponding port after all initialization is finished. In our example, an event inserted into the timerStart port starts the HWClock, and the rest of the execution is driven by clock interrupt events.

The TinyGALS programming model has the advantage that actors become decoupled through message passing and are easy to develop independently. However, each message passed will trigger the scheduler and activate a receiving actor, which may quickly become inefficient if there is a global state that must be shared among multiple actors. TinyGUYS (Guarded Yet Synchronous) variables are a mechanism for sharing global state, allowing quick access but with protected modification of the data.

In the TinyGUYS mechanism, global variables are guarded. Actors may read the global variables synchronously (without delay). However, writes to the variables are asynchronous in the sense that all writes are buffered. The buffer is of size one, so the last actor that writes to a variable wins. TinyGUYS variables are updated by the scheduler only when it is safe (e.g., after one module finishes and before the scheduler triggers the next module).

TinyGUYS have global names defined at the application level which are mapped to the parameters of each actor and are further mapped to the external variables of the components that use these variables. The external variables are accessed within a component by using special keywords: PARAM_GET and PARAM_PUT. The code generator produces thread-safe implementation of these methods using locking mechanisms, such as turning off interrupts.

TinyGALS Code Generation

TinyGALS takes a generative approach to mapping high-level constructs such as FIFO queues and actors into executables on Berkeley motes. Given the highly structured architecture of TinyGALS applications, efficient scheduling and event handling code can be automatically generated to free software developers from writing error-prone concurrency control code. The rest of this section discusses a code generation tool that is implemented based on TinyOS v0.6.1 for Berkeley motes.

Given the definitions for the components, actors, and application, the code generator automatically generates all of the necessary code for (1) component links and actor connections, (2) application initialization and start of execution, (3) communication among actors, and (4) global variable reads and writes.

Similar to how TinyOS deals with connected method calls among components, the TinyGALS code generator generates a set of aliases for each synchronous method call. The code generator also creates a system-level initialization function called app_init(), which contains calls to the init() method of each actor in the system. The app_init() function is one of the first functions called by the TinyGALS run-time scheduler before executing the application. An application start function app_start() is created based on the @start annotation. This function triggers the input port of the actor defined as the application starting point.

The code generator automatically generates a set of scheduler data structures and functions for each asynchronous connection between actors. For each input port of an actor, the code generator generates a queue of length n, where n is specified in the application definition. The width of the queue depends on the number of arguments of the method connected to the port. If there are no arguments, then as an optimization, no queue is generated for the port (but space is still reserved for events in the scheduler event queue).

For each output port of an actor, the code generator generates a function that has the same name as the output port. This function is called whenever a method of a component wishes to write to an output port. The type signature of the output port function matches that of the method that connects to the port. For each input port connected to the output port, a put() function is generated which handles the actual copying of data to the input port queue. The output port function calls the input port's put() function for each connected input port. The put() function adds the port identifier to the scheduler event queue so that the scheduler will activate the actor at a later time.

For each connection between a component method and an actor input port, a function is generated with a name formed from the name of the input port and the name of the component method. When the scheduler activates an actor via an input port, it first calls this generated function to remove data from the input port queue and then passes it to the component method.

For each TinyGUYS variable declared in the application definition, a pair of data structures and a pair of access functions are generated. The pair of data structures consists of a data storage location of the type specified in the module definition that uses the global variable, along with a buffer for the storage location. The pair of access functions consists of a PARAM_GET() function that returns the value of the global variable, and a PARAM_PUT() function that stores a new value for the variable in the variable's buffer. A generated flag indicates whether the scheduler needs to update the variables by copying data from the buffer.

Since most of the data structures in the TinyGALS run-time scheduler are generated, the scheduler does not need to worry about handling different data types and the conversion among them. What is left in the run-time scheduler is merely event-queuing and function-triggering mechanisms. As a result, the TinyGALS run-time scheduler is very lightweight. The scheduler itself takes 112 bytes of memory, comparable with the original 86-byte TinyOS v0.6.1 scheduler.

7.4 Node-Level Simulators

Node-level design methodologies are usually associated with simulators that simulate the behavior of a sensor network on a per-node basis. Using simulation, designers can quickly study the performance (in terms of timing, power, bandwidth, and scalability) of potential algorithms without implementing them on actual hardware and dealing with the vagaries of actual physical phenomena.

A node-level simulator typically has the following components:

- *Sensor node model:* A node in a simulator acts as a software execution platform, a sensor host, as well as a communication terminal. In order for designers to focus on the application-level code, a node model typically provides or simulates a communication protocol stack, sensor behaviors (e.g., sensing noise), and operating system services. If the nodes are mobile, then the positions and motion properties of the nodes need to be modeled. If energy characteristics are part of the design considerations, then the power consumption of the nodes needs to be modeled.

- *Communication model:* Depending on the details of modeling, communication may be captured at different layers. The most elaborate simulators model the communication media at the physical layer, simulating the RF propagation delay and collision of simultaneous transmissions. Alternately, the communication may be simulated at the MAC layer or network layer, using, for example, stochastic processes to represent low-level behaviors.

- *Physical environment model:* A key element of the environment within which a sensor network operates is the physical phenomenon of interest. The environment can also be simulated at various levels of detail. For example, a moving object in the physical world may be abstracted into a point signal source. The motion of the point signal source may be modeled by differential equations or interpolated from a trajectory profile. If the sensor network is passive—that is, it does not impact the behavior of the environment—then the environment can be simulated separately or can even be stored in data files for sensor nodes to read in. If, in addition to sensing, the network also performs actions that influence the behavior of the environment, then a more tightly integrated simulation mechanism is required.

- *Statistics and visualization:* The simulation results need to be collected for analysis. Since the goal of a simulation is typically to derive global properties from the execution of individual nodes, visualizing global behaviors is extremely important. An ideal visualization tool should allow users to easily observe on demand the spatial distribution and mobility of the nodes, the connectivity among nodes, link qualities, end-to-end communication routes and delays, phenomena and their spatio-temporal dynamics, sensor readings on each node, sensor node states, and node lifetime parameters (e.g., battery power).

A sensor network simulator simulates the behavior of a subset of the sensor nodes with respect to time. Depending on how the time is advanced in the simulation, there are two types of execution models: *cycle-driven simulation* and *discrete-event simulation*. A cycle-driven (CD) simulation discretizes the continuous notion of real time into (typically regularly spaced) ticks and simulates the system behavior at these ticks. At each tick, the physical phenomena are first simulated, and then all nodes are checked to see if they have anything to sense, process, or communicate. Sensing and computation are assumed to be finished before the next tick. Sending a packet is also assumed to be completed by then. However, the packet will not be available for the destination node until the next tick. This split-phase communication is a key mechanism to reduce cyclic dependencies that may occur in cycle-driven simulations. That is, there should be no two components, such that one of them computes $y_k = f(x_k)$ and the other computes $x_k = g(y_k)$, for the same tick index k. In fact, one of the most subtle issues in designing a CD simulator is how to detect and deal with cyclic dependencies among nodes or algorithm components. Most CD simulators do not allow interdependencies within a single tick. Synchronous languages [91], which are typically used in control system designs rather than sensor network designs, do allow cyclic dependencies. They use a fixed-point semantics to define the behavior of a system at each tick.

Unlike cycle-driven simulators, a discrete-event (DE) simulator assumes that the time is continuous and an event may occur at any time. An event is a 2-tuple with a value and a time stamp indicating when the event is supposed to be handled. Components in a DE simulation react to input events and produce output events. In node-level simulators, a component can be a sensor node and the events can be communication packets; or a component can be a software module within a node and the events can be message passings among these modules. Typically, components are *causal*, in the sense that if an output event is computed from an input event, then the time stamp of the output event should not be earlier than that of the input event. Noncausal components require the simulators to be able to roll back in time, and, worse, they may not define a deterministic behavior of a system [129]. A DE simulator typically requires a global event queue. All events passing between nodes or modules are put in the event queue and sorted according to their chronological order. At each iteration of the simulation, the simulator removes the first event (the one with the earliest time stamp) from the queue and triggers the component that reacts to that event.

In terms of timing behavior, a DE simulator is more accurate than a CD simulator, and, as a consequence, DE simulators run slower. The overhead of ordering all events and computation, in addition to the values and time stamps of events, usually dominates the computation time. At an early stage of a design when only the asymptotic behaviors rather than timing properties are of concern, CD simulations usually require less complex components and give faster simulations. Partly because of the approximate timing behaviors, which make simulation results less comparable from application to application, there is no general CD simulator that fits all sensor network simulation tasks. We have come across a number of home-grown simulators written in Matlab, Java, and C++. Many of them are developed for particular applications and exploit application-specific assumptions to gain efficiency.

DE simulations are sometimes considered as good as actual implementations, because of their continuous notion of time and discrete notion of events. There are several open-source or commercial simulators available. One class of these simulators comprises extensions of classical network simulators, such as ns-2,[6] J-Sim (previously known as JavaSim),[7] and GloMoSim/QualNet.[8] The focus of these simulators is on network modeling, protocols stacks, and simulation performance. Another class of simulators, sometimes called *software-in-the-loop simulators*, incorporate the actual node software into the simulation. For this reason, they are typically attached to particular hardware platforms and are less portable. Examples include TOSSIM [131] for Berkeley motes and Em* (pronounced *em star*) [62] for Linux-based nodes such as Sensoria WINS NG platforms.

7.4.1 The ns-2 Simulator and its Sensor Network Extensions

The simulator ns-2 is an open-source network simulator that was originally designed for wired, IP networks. Extensions have been made

---

6 Available at *http://www.isi.edu/nsnam/ns*.
7 Available at *http://www.j-sim.org*.
8 Available at *http://pcl.cs.ucla.edu/projects/glomosim*.

to simulate wireless/mobile networks (e.g., 802.11 MAC and TDMA MAC) and more recently sensor networks. While the original ns-2 only supports logical addresses for each node, the wireless/mobile extension of it (e.g., [25]) introduces the notion of node locations and a simple wireless channel model. This is not a trivial extension, since once the nodes move, the simulator needs to check for each physical layer event whether the destination node is within the communication range. For a large network, this significantly slows down the simulation speed.

There are at least two efforts to extend ns-2 to simulate sensor networks: SensorSim from UCLA[9] and the NRL sensor network extension from the Navy Research Laboratory.[10] SensorSim aims at providing an energy model for sensor nodes and communication, so that power properties can be simulated [175]. SensorSim also supports hybrid simulation, where some real sensor nodes, running real applications, can be executed together with a simulation. The NRL sensor network extension provides a flexible way of modeling physical phenomena in a discrete event simulator. Physical phenomena are modeled as network nodes which communicate with real nodes through physical layers. Any interesting events are sent to the nodes that can sense them as a form of communication. The receiving nodes simply have a sensor stack parallel to the network stack that processes these events.

The main functionality of ns-2 is implemented in C++, while the dynamics of the simulation (e.g., time-dependent application characteristics) is controlled by Tcl scripts. Basic components in ns-2 are the layers in the protocol stack. They implement the *handlers* interface, indicating that they handle events. Events are communication packets that are passed between consecutive layers within one node, or between the same layers across nodes.

The key advantage of ns-2 is its rich libraries of protocols for nearly all network layers and for many routing mechanisms. These protocols

---

9 Available at *http://nesl.ee.ucla.edu/projects/sensorsim/*.

10 Available at *http://pf.itd.nrl.navy.mil/projects/nrlsensorsim/*.

are modeled in fair detail, so that they closely resemble the actual protocol implementations. Examples include the following:

- TCP: reno, tahoe, vegas, and SACK implementations

- MAC: 802.3, 802.11, and TDMA

- Ad hoc routing: Destination sequenced distance vector (DSDV) routing, dynamic source routing (DSR), ad hoc on-demand distance vector (AODV) routing, and temporally ordered routing algorithm (TORA)

- Sensor network routing: Directed diffusion, geographical routing (GEAR) and geographical adaptive fidelity (GAF) routing.

7.4.2 The Simulator TOSSIM

TOSSIM is a dedicated simulator for TinyOS applications running on one or more Berkeley motes. The key design decisions on building TOSSIM were to make it scalable to a network of potentially thousands of nodes, and to be able to use the actual software code in the simulation. To achieve these goals, TOSSIM takes a cross-compilation approach that compiles the nesC source code into components in the simulation. The event-driven execution model of TinyOS greatly simplifies the design of TOSSIM. By replacing a few low-level components, such as the A/D conversion (ADC), the system clock, and the radio front end, TOSSIM translates hardware interrupts into discrete-event simulator events. The simulator event queue delivers the interrupts that drive the execution of a node. The upper-layer TinyOS code runs unchanged.

TOSSIM uses a simple but powerful abstraction to model a wireless network. A network is a *directed* graph, where each vertex is a sensor node and each directed edge has a bit-error rate. Each node has a private piece of state representing what it hears on the radio channel. By setting connections among the vertices in the graph and a bit-error rate on each connection, wireless channel characteristics, such as imperfect channels, hidden terminal problems, and asymmetric links, can be easily modeled. Wireless transmissions are simulated at the bit level. If a bit error occurs, the simulator flips the bit.

TOSSIM has a visualization package called TinyViz, which is a Java application that can connect to TOSSIM simulations. TinyViz also provides mechanisms to control a running simulation by, for example, modifying ADC readings, changing channel properties, and injecting packets. TinyViz is designed as a communication service that interacts with the TOSSIM event queue. The exact visual interface takes the form of plug-ins that can interpret TOSSIM events. Beside the default visual interfaces, users can add application-specific ones easily.

7.5 Programming Beyond Individual Nodes: State-Centric Programming

Many sensor network applications, such as target tracking, are not simply generic distributed programs over an ad hoc network of energy-constrained nodes. Deeply rooted in these applications is the notion of states of physical phenomena and models of their evolution over space and time. Some of these states may be represented on a small number of nodes and evolve over time, as in the target tracking problem in Chapter 2, while others may be represented over a large and spatially distributed number of nodes, as in tracking a temperature contour.

A distinctive property of physical states, such as location, shape, and motion of objects, is their continuity in space and time. Their sensing and control is typically done through sequential state updates. System theories, the basis for most signal and information processing algorithms, provide abstractions for state update, such as:

$$\mathbf{x}_{k+1} = f(\mathbf{x}_k, u_k) \qquad (7.1)$$

$$\mathbf{y}_k = g(\mathbf{x}_k, u_k) \qquad (7.2)$$

where $\mathbf{x}$ is the state of a system, u are the inputs, $\mathbf{y}$ are the outputs, k is an integer update index over space and/or time, f is the state update function, and g is the output or observation function. This formulation is broad enough to capture a wide variety of algorithms in sensor fusion, signal processing, and control (e.g., Kalman filtering, Bayesian estimation, system identification, feedback control laws, and finite-state automata).

However, in distributed real-time embedded systems such as sensor networks, the formulation is not so clean as represented in those equations. The relationships among subsystems can be highly complex and dynamic over space and time. The following concerns, not explicitly raised (7.1) and (7.2), must be properly addressed during the design to ensure the correctness and efficiency of the resulting systems.

- Where are the state variables stored?

- Where do the inputs come from?

- Where do the outputs go?

- Where are the functions f and g evaluated?

- How long does the acquisition of inputs take?

- Are the inputs in u_k collected synchronously?

- Do the inputs arrive in the correct order through communication?

- What is the time duration between indices k and $k + 1$? Is it a constant?

These issues, addressing *where* and *when*, rather than *how*, to perform sensing, computation, and communication, play a central role in the overall system performance. However, these "nonfunctional" aspects of computation, related to concurrency, responsiveness, networking, and resource management, are not well supported by traditional programming models and languages. State-centric programming aims at providing design methodologies and frameworks that give meaningful abstractions for these issues, so that system designers can continue to write algorithms like (7.1) and (7.2) on top of an intuitive understanding of where and when the operations are performed. This section introduces one such abstraction, namely, collaboration groups.

7.5.1 Collaboration Groups

A collaboration group is a set of entities that contribute to a state update. These entities can be physical sensor nodes, or they can be more abstract system components such as virtual sensors or mobile agents hopping among sensors. In this context, they are all referred to as *agents*.

Intuitively, a collaboration group provides two abstractions: its *scope* to encapsulate network topologies and its *structure* to encapsulate communication protocols. The scope of a group defines the membership of the nodes with respect to the group. For the discussion of collaboration groups in this chapter, we broaden the notion of nodes to include both physical sensor nodes and virtual sensor nodes that may not be attached to any physical sensor. In this broader sense of node, a software agent that hops among the sensor nodes to track a target is a virtual node. Limiting the scope of a group to a subset of the entire space of all agents improves scalability. The scope of a group can be specified existentially or by a membership function (e.g., all nodes in a geometric extent, all nodes within a certain number of hops from an anchor node, or all nodes that are "close enough" to a temperature contour). Grouping nodes according to some physical attributes rather than node addresses is an important and distinguishing characteristic of sensor networks.

The *structure* of a group defines the "roles" each member plays in the group, and thus the flow of data. Are all members in the group equal peers? Is there a "leader" member in the group that consumes data? Do members in the group form a tree with parent and children relations? For example, a group may have a leader node that collects certain sensor readings from all followers. By mapping the leader and the followers onto concrete sensor nodes, we effectively define the flow of data from the hosts of followers to the host of the leader. The notion of roles also shields programmers from addressing individual nodes either by name or address. Furthermore, having multiple members with the same role provides some degree of redundancy and improves robustness of the application in the presence of node and link failures.

Formally, a group is a 4-tuple:

$$G = (A, L, p, R)$$

where

A is a set of agents;

L is a set of labels, called *roles*;

$p : A \rightarrow L$ is a function that assigns each agent a role;

$R \subseteq L \times L$ are the connectivity relations among roles.

Given the relations among roles, a group can induce a lower-level connectivity relation E among the agents, so that $\forall a, b \in A$, if $(p(a), p(b)) \in R$, then $(a, b) \in E$. For example, under this formulation, the leader-follower structure defines two roles, $L = \{leader, follower\}$, and a connectivity relation, $R = \{(follower, leader)\}$, meaning that the follower sends data to the leader. Then, by specifying one leader agent and multiple follower agents within a geographical region (i.e., specifying a map p from a set of agents in A to labels in L), we have effectively specified that all followers send data to the leader without addressing the followers individually.

At run time, the scope and structural dynamics of groups are managed by group management protocols, which are highly dependent on the types of groups. A detailed specification of group management protocols is beyond the scope of this section. Some examples of these protocols are discussed here at a high level. Interested readers can refer to Chapter 3 for more detail.

Examples of Groups
Combinations of scopes and structures create patterns of groups that may be highly reusable from application to application. Here, we give

several examples of groups, though by no means is it a complete list. The goal is to illustrate the wide variety of the kinds of groups, and the importance of mixing and matching them in applications.

Geographically Constrained Group. A geographically constrained group (GCG) consists of members within a prespecified geographical extent. Since physical signals, especially the ones from point targets, may propagate only to a limited extent in an environment, this kind of group naturally represents all the sensor nodes that can possibly "sense" a phenomenon. There are many ways to specify the geographic shape, such as circles, polygons, and their unions and intersections. A GCG can be easily established by geographically constrained flooding. Protocols such as Geocasting [117], GEAR [229], and Mobicast [102] may be used to support the communication among members even in the presence of communication "holes" in the region. A GCG may have a leader, which fuses information from all other members in the group.

N-hop Neighborhood Group. When the communication topology is more important than the geographical extent, hop counts are useful to constrain group membership. An *n*-hop neighborhood group (*n*-HNG) has an anchor node and defines that all nodes within *n* communication hops are members of the group. Since it uses hop counts rather than Euclidean distances, local broadcasting can be used to determine the scope. Usually, the anchor node is the leader of the group, and the group may have a tree structure with the leader as the root to optimize for communication. If the leader's behavior can be decomposed into suboperations running on each node, then the tree structure also provides a platform for distributing the computation.

There are several useful special cases for *n*-HNG. For example, 0-HNG contains only the anchor node itself, 1-HNG comprises the one-hop neighbors of the anchor node, and ∞-HNG contains all the nodes reachable from the root. From this point of view, TinyDB [149] (as discussed in Chapter 6) is built on a ∞-HNG group.

Publish/Subscribe Group. A group may also be defined more dynamically, by all entities that can provide certain data or services, or that can satisfy certain predicates over their observations or internal states. A publish/subscribe group (PSG) comprises consumers expressing interest in specific types of data or services and producers that provide those data or services. Communication among members of a PSG may be established via rendezvous points, directory servers, or network protocols such as directed diffusion.

Acquaintance Group. An even more dynamic kind of group is the acquaintance group (AG), where a member belongs to the group because it was "invited" by another member in the group. The relationships among the members may not depend on any physical properties at the current time but may be purely logical and historical. A member may also quit the group without requiring permission from any other member. An AG may have a leader, serving as the rendezvous point. When the leader is also fixed on a node or in a region, GPSR [112], ad hoc routing trees, or directed diffusion types of protocols may facilitate the communication between the leader and the other members. An obvious use of this group is to monitor and control mobile agents from a base station. When all members in the group are mobile, there is no leader member, and any member may wish to communicate to one or more other members, the maintenance of connectivity among the group members can be nontrivial. The roaming hub (RoamHBA) protocol is an example of maintaining connectivity among mobile agents [67].

Using Multiple Types of Groups

Mixing and matching groups is a powerful technique for tackling system complexity by making algorithms much more scalable and resource efficient without sacrificing conceptual clarity. One may use highly tuned communication protocols for specific groups to reduce latency and energy costs.

There are various ways to compose groups. They can be composed in parallel to provide different types of input for a single computational entity. For example, in the target tracking problem in

Chapters 2 and 5, one may use a GCG to gather sensor measurements, while using a 1-HNG to select the potential next leader. Groups may also be composed hierarchically, such that a group (or its representative member) is contained by another group. For example, while using multiple groups to compute target trajectories, all tracking leaders of various targets may form a PSG with a base station to report the tracking result to.

7.5.2 PIECES: A State-Centric Design Framework

PIECES (Programming and Interaction Environment for Collaborative Embedded Systems) [141] is a software framework that implements the methodology of state-centric programming over collaboration groups to support the modeling, simulation, and design of sensor network applications. It is implemented in a mixed Java-Matlab environment.

Principals and Port Agents

PIECES comprises *principals* and *port agents*. Figure 7.16 shows the basic relations among principals and port agents.

A principal is the key component for maintaining a piece of *state*. Typically, a principal maintains state corresponding to certain aspects of the physical phenomenon of interest.[11] The role of a principal is to update its state from time to time, a computation corresponding to evaluating function f in (7.1). A principal also accepts other principals' queries of certain views on its own state, a computation corresponding to evaluating function g in (7.2).

To update its portion of the state, a principal may gather information from other principals. To achieve this, a principal creates port agents and attaches them onto itself and onto the other principals. A port agent may be an input, an output, or both. An output port

---

11 From a computational perspective, a port agent as an object certainly has its own state. But the distinction here is that the states of port agents are *not* about physical phenomena.

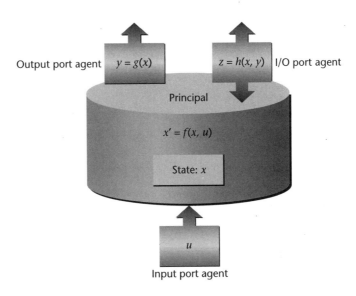

Output port agent $y = g(x)$

$z = h(x, y)$ I/O port agent

Principal

$x' = f(x, u)$

State: x

u

Input port agent

Figure 7.16 Principal and port agents (adapted from [141]).

agent is also called an *observer*, since it computes outputs based on the host principal's state and sends them to other agents. Observers may be active or passive. An active observer pushes data autonomously to its destination(s), while a passive observer sends data only when a consumer requests it. A principal typically attaches a set of observers to other principals and creates a local input port agent to receive the information collected by the remote agents. Thus port agents capture communication patterns among principals.

The execution of principals and port agents can be either time-driven or event-driven, where events may include physical events that are pushed to them (i.e., data-driven) or query events from other principals or agents (i.e., demand-driven). Principals maintain state, reflecting the physical phenomena. These states can be updated, rather than rediscovered, because the underlying physical states are typically continuous in time. How often the principal states need to be updated depends on the dynamics of the phenomena or physical events. The executions of observers, however, reflect the demands of the outputs. If an output is not currently needed, there is no need to compute it. The notion of "state" effectively separates these two execution flows.

To ensure consistency of state update over a distributed computational platform, PIECES requires that a piece of state, say $\mathbf{x}|_s$, can only be maintained by exactly one principal. Note that this does not prevent other principals from having local caches of $\mathbf{x}|_s$ for efficiency and performance reasons; nor does it prevent the other principals from locally updating the values of cached $\mathbf{x}|_s$. However, there is only one "master copy" for $\mathbf{x}|_s$; all local updates should be treated as "suggestions" to the master copy, and only the principal that owns $\mathbf{x}|_s$ has the final word on its values. This asymmetric access of variables simplifies the way shared variables are managed.

Principal Groups

Principals can form groups. A principal group gives its members a means to find other relevant principals and attaches port agents to them. A principal may belong to multiple groups. A port agent, however, serving as a proxy for a principal in the group, can only be associated with one group.

The creation of groups can be delegated to port agents, especially for leader-based groups. The leader port agent, typically of type input, can be created on a principal, and the port agent can take group scope and structure parameters to find the other principals and create follower port agents on them. Groups can be created dynamically, based on the collaboration needs of principals. For example, when a tracking principal finds that there is more than one target in its sensing region, it may create a classification group to fulfill the need of classifying the targets. A group may have a limited time span. When certain collaborations are no longer needed, their corresponding groups can be deleted.

The structure of a group allows its members to address other principals through their role, rather than their name or logical address. For example, the only interface that a follower port agent in a leader-follower structured group needs is to send data to the leader. If the leader moves to another node while a data packet is moving from a follower agent to the leader, the group management protocol should take care of the dangling packet, either delivering it to the leader at the new location or simply discarding it. The group management protocol may be built on top of data-centric routing and storage services such as diffusion routing and GHT (discussed in earlier chapters).

Mobility

A principal is hosted by a specific network node at any given time. The most primitive type of principal is a *sensing principal*, which is fixed to a sensor node. A sensing principal maintains a piece of (local) state related to the physical phenomenon, based solely on its own local measurement history. Although a sensing principal is constrained to a physical node, other principals may be implemented as software agents that move from host to host, depending on information utility, performance requirements, time constraints, and resource availability. A principal P may also be *attached* to another principal Q in the sense that P moves with Q. When a principal moves, it carries its state to the new location and the scope of the group it belongs to may be updated if necessary.

Mobile principals bring additional challenges to maintaining the state. For example, a principal should not move while it is in the middle of updating the state. To ensure this, PIECES imposes the restriction that whenever an agent is triggered, its execution must have reached a quiescent state. Such a trigger is called a *responsible trigger* [147]. Only at these quiescent states can principals move to other nodes in a well-defined way, carrying a minimum amount of information representing the phenomena.

PIECES Simulator

PIECES provides a mixed-signal simulator that simulates sensor network applications at a high level. The simulator is implemented using a combination of Java and Matlab. An event-driven engine is built in Java to simulate network message passing and agent execution at the collaboration-group level. A continuous-time engine is built in Matlab to simulate target trajectories, signals and noise, and sensor front ends. The main control flow is in Java, which maintains the global notion of time. The interface between Java and Matlab also makes it possible to implement functional algorithms such as signal processing and sensor fusion in Matlab, while leaving their execution control in Java. A three-tier distributed architecture is designed through Java registrar and RMI interfaces, so that the execution in Java and Matlab can be separately interrupted and debugged.

Like most network simulators such as ns-2, the PIECES simulator maintains a global event queue and triggers computational entities—principals, port agents, and groups—via timed events. However, unlike network simulators that aim to accurately simulate network behavior at the packet level, the PIECES simulator verifies CSIP algorithms in a networked execution environment at the collaboration-group level. Although groups must have distributed implementations in real deployments, they are centralized objects in the simulator. They can internally make use of instant access to any member of any role, although these services are not available to either principals or port agents. This relieves the burden of having to develop, optimize, and test the communication protocols concurrently with the CSIP algorithms. The communication delay is estimated based on the locations of sender and receiver and the group management protocol being used. For example, if an output port of a sensing principal calls sendToLeader(message) on its container group, then the group determines the sensor nodes that host the sensing principal and the destination principal, computes the number of hops between the two nodes specified by the group management protocol, and generates a corresponding delay and a bit error based on the number of hops. A detailed example of using this simulator is given in the next section.

7.5.3   Multitarget Tracking Problem Revisited

Using the state-centric model, programmers decouple a global state into a set of independently maintained pieces, each of which is assigned a principal. To update the state, principals may look for inputs from other principals, with sensing principals supporting the lowest-level sensing and estimation tasks. Communication patterns are specified by defining collaboration groups over principals and assigning corresponding roles for each principal through port agents. A mobile principal may define a utility function, to be evaluated at candidate sensor nodes, and then move to the best next location, all in a way transparent to the application developer. Developers can focus on implementing the state update functions as if they are writing centralized programs.

To make these concepts concrete, let us revisit the multitarget tracking system introduced in Chapter 2. Recall that in Figure 2.5, the tracking of two crossing targets can be decomposed into three phases:

1. When the targets are far apart, the tracking problem can be treated as a set of single-target tracking subproblems.

2. When the targets are in proximity of each other, they are tracked jointly due to signal mixing.

3. After the targets move apart, the tracking problem becomes two single-target tracking subproblems again.

To summarize, there are two kinds of target information that the user cares about in this context: target positions and target identities. In the third phase above, in addition to the problem of updating track locations, there is a need to sort out ambiguity regarding which track corresponds to which target. We refer to this problem as the *identity management* problem. Specifically, one must keep track of how the identities mix when targets cross over, and update identity information at the other node when credible target identity evidence is available to one node. The identity information may be obtained by a local classifier or by an identity management protocol across tracks. In PIECES, the system is designed as a set of communicating target trackers (MTTrackers), where each tracker maintains the trajectory and identity information about a target or a set of spatially adjacent targets. An MTTracker is implemented by three principals: a *tracking principal*, a *classification principal*, and an *identity management principal*, as shown in Figure 7.17. In the first phase, the identity state of the track is trivial; thus no classification and identity management principals are needed.

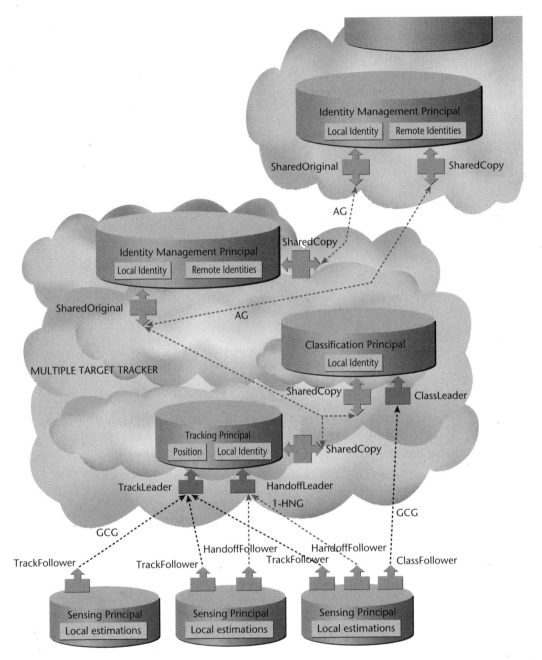

Figure 7.17 The distributed multi-object tracking algorithm as implemented in the state-centric programming model, using distributed principals and agents as discussed in the text. Notice that the state-centric model allows an application developer to focus on key pieces of state information the sensor network creates and maintains, thus raising the abstraction level of programming (adapted from [141]).

A tracking principal updates the track position state periodically. It collects local individual position estimates from sensors close to the target by a GCG with a leader-follower relation. The tracking principal is the leader, and all sensing principals within a certain geographical extent centered about the current target position estimate are the followers. The tracking principal also makes hopping decisions based on its current position estimate and the node characteristic information collected from its one-hop neighbors via a 1-HNG. When the principal is initialized, it creates the agents and corresponding groups. Behind the scene, the groups create follower agents with specific types of output, indicated by the sensor modalities. Without further instructions from the programmer, the followers periodically report their outputs to the input port agents. Whenever the leader principal is activated by a time trigger, it updates the target position using the newly received data from the followers and selects the next hosting node based on neighbor node characteristics.

Both the classification principal and the identity management principal operate on the identity state, with the identity management principal maintaining the "master copy" of the state. In fact, the classification principal is created only when there is a need for classifying targets. The classification principal uses a GCG to collect class feature information from nearby sensing principals in the same way that tracking principals collect location estimates. The identity management principal forms an AG with all other identity management principals that may have relevant identity information. They become members of a particular identity group only when targets intersect and their identities mix. Both classification principals and identity management principals are *attached* to the tracking principal for their mobility decisions. However, the formation of an AG among these three principals also provides the flexibility that they can make their own hopping decisions without changing their interaction interface.

Simulation Results

Figure 7.18 shows the progression of tracking two crossing targets. Initially, when the targets are well separated, as in Figure 7.18(a),

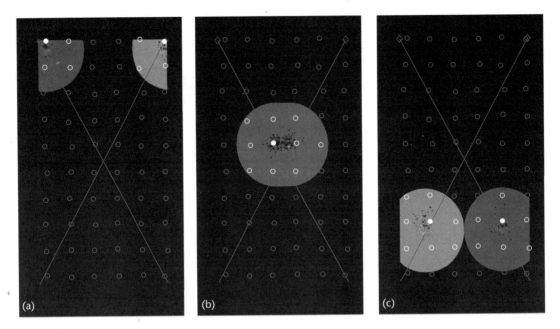

Figure 7.18 Simulation snapshots: Sensor nodes are indicated by small circles, and the crossing lines indicate the true trajectories of the two targets. One geographically constrained group is created for each target. When the two targets cross over, their groups merge into one.

each target is tracked by a tracker whose sensing group is pictured as a shaded disk. The hosting node of the tracking principal is plotted in solid white dots, and the hosts for corresponding sensing principals are plotted in small, empty white circles inside the shaded disks. Since the targets are well separated, each identity group contains only one member—the identity management principal of a tracker. As the targets move toward the center of the sensor field, the sensing groups move with their respective track positions. In Figure 7.18(b), the two separate tracking groups have merged. A joint tracking principal updates tracks for both targets. The reason for the merge is that when the two targets approach each other, it is more accurate to track the targets jointly, rather than independently, due to the effect of signal mixing. Finally, as the targets move away from each other, the merged tracking group splits into two separate single-target tracking groups that proceed to track each target separately, as shown in Figure 7.18(c). At this point, the identities of the targets are mixed, so that an identity group is created to contain the two identity management principals from both trackers.

Figure 7.19 shows a snapshot of a more complicated multitarget crossover scenario. Three tracks (A, B, and D) have crossed one another at some point in time; hence the identities of these tracks are mixed. The corresponding identity management principals form an identity group. Now, one identity management principal (the one at the bottom-right corner) collects classification information and identifies the track as belonging to the red target, as shown in the figure. Hence, it communicates with its peers in the identity management group (top-right and bottom-middle principals) to update the identity of their respective targets as well. The updated identity is shown in the right-hand-side bar chart in Figure 7.19. Defining the acquaintance group and its interface in this way allows these spatially distributed identity management principals to communicate with one another, thus providing the application developer the necessary abstraction for focusing on the functional aspect of identity management algorithms without worrying about the communication details.

7.6 Summary

This chapter has provided an overview of sensor network hardware and software platforms and application design methodologies. Although most of the existing platforms are tightly bound to particular hardware designs, the design principles covered in this chapter can be generalized to other hardware platforms as well. We described TinyOS, nesC, and TinyGALS, as examples of node-level operating systems and programming languages based on the Berkeley mote hardware. The node-centric platforms typically employ a message-passing abstraction of communication between nodes. Several networking protocols are covered in Chapter 3. Interfaces

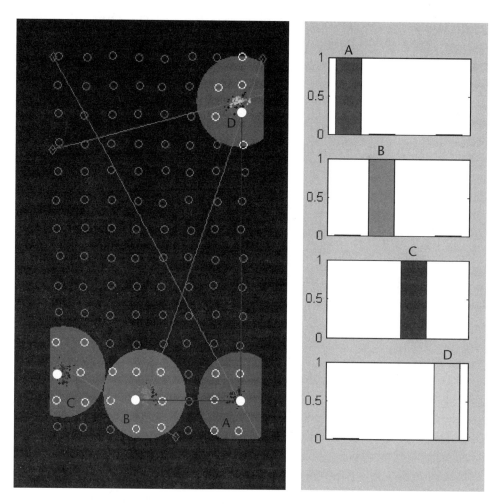

Figure 7.19 Acquaintance group for identity management: The left panel shows the tracker group and tracking results, and the right panel shows the identity of each track as bar charts. On the right panel, each subfigure represents the identity belief of each tracker. For example, in the first subfigure, the identity management principal—the one near the bottom right on the left panel—believes that the track it is maintaining is most likely that of the target A, with a much smaller probability of being target B or target D.

for these protocols include the active messaging of TinyOS and the publish/subscribe interface of directed diffusion.

Programming distributed systems beyond individual nodes has been traditionally handled by middleware technologies such as CORBA [171]. However, these are not directly suitable for ad hoc resource-constrained sensor networks. State-centric programming is aimed at providing domain-specific programming models for information processing applications in sensor networks. This allows programmers to think in terms of high-level abstractions such as the state of the physical phenomena of interest, rather than the behaviors of individual nodes. Ultimately, in a dense sensor network, the behavior of each individual node is not as important, and sensor network applications should be robust to individual node failures. It is the collective behavior that makes sensor networks viable and interesting.

Programming methodologies and tools for sensor networks is a new area of research, as new network organization principles and new programming abstractions emerge and as more sensor network applications are built. Just as hardware-description languages (e.g., VHDL and Verilog) have served the VLSI designs, and as control languages (e.g., Signal [128], Esterel [17], and Giotto [97]) and architectures (e.g., time-triggered architecture [119]) have supported real-time control system designs, domain-specific programming models and tools will be key to the development and deployment of large-scale sensor network applications.

8

Applications and Future Directions

We have covered many important topics on designing and deploying large-scale sensor networks in the previous chapters. Our primary focus has been the information processing aspects of sensor networks—namely, how sensor information is acquired, represented, processed, transmitted, aggregated, and accessed. This chapter will summarize the salient points in these discussions, describe emerging sensor network applications, and outline future research directions.

8.1 A Summary of the Book

We have advocated a holistic approach to the design of sensor networks, including optimizing across physical layers, networking, embedded OS, node services, group management, collaborative processing, and application-specific needs. Figure 8.1 shows one possible organization of a sensor network stack. While the exact layers and interactions are still debatable and depend on the particular system requirements and environmental constraints, a sensor network stack must at least support information processing *across* multiple nodes in a resource-aware manner. Hence, the central theme of what we have presented so far can be summarized as developing scalable algorithms and software architectures to support distributed information processing applications on resource-constrained sensor networks.

To describe the key elements of a typical sensor network system, we introduced distributed object localization and tracking as a

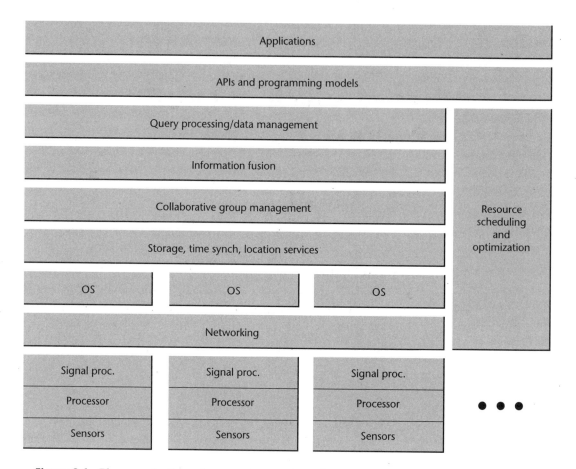

Figure 8.1   The organization of a sensor network stack.

canonical problem (Chapter 2). We studied a number of information estimation and fusion techniques that are well suited for distributed signal processing and discussed the information processing and communication requirements for the tracking problem. To meet these requirements, we presented techniques such as geographic and attribute-based routing for sensor networking (Chapter 3). The key idea was to name and route data based on physical attributes such as location rather than logical (IP type) node addresses. We investigated a number of techniques for establishing a common time base across nodes and for nodes to discover their locations (Chapter 4); these are important services that enable networks to initialize themselves and carry out collaborative sensing and processing tasks.

Sensor tasking and control is an important topic for resource-constrained sensor networks. We introduced a number of information-driven approaches that optimize for information gain and resource utilization by solving a distributed constrained optimization problem (Chapter 5). Intrinsic to these techniques is the notion of collaborative group management that aids the formation of dynamic, ad hoc groups of sensors to support distributed processing tasks. Treating a sensor network as a distributed database, we studied the problem of storing, locating, and accessing data in a distributed, power-aware setting (Chapter 6). We discussed the topics of in-network aggregation, data-centric storage, range search, and multiresolution summarization. Finally, we studied the problem of system architectures and programming models for sensor network applications (Chapter 7). Because of the cross-node interactions and event-driven nature of many sensing applications, appropriate software architectures have to effectively manage concurrency and resource allocation, while presenting to application developers a model of programming that is closer to application semantics than to lower-level system programming.

An important lesson we can draw from these studies is the unavoidable interleaving of information processing and software architecture in sensor network systems, and the need to co-design and co-optimize these two architectures simultaneously. For example, in the IDSQ algorithm we discussed (Chapter 5), the information flow from node to node, as dictated by the Bayesian probabilistic tracker, requires the corresponding networking protocols, group management, and software frameworks to support this style of information processing and communication. As we gain more experience with a broad range of sensing applications, we can expect to discover other important patterns in the interactions between the information processing and software architectures, and design algorithms and tools to exploit these patterns.

8.2 Emerging Applications

Although 99 percent of today's sensors are still wired, wireless sensors offer significant advantages over wired sensors. The main problem with wired sensor networks is cost and delays in deployment. It is estimated that the dominant factor in implementing wired sensing systems is the cost of wiring, which ranges from $40 to $2000 per linear foot of wire. A wired network is also more time consuming to construct and deploy, precluding applications where immediate data collection is needed. This makes infeasible certain applications, such as automotive or battlefield sensing, where mobility and rapid deployment are of the essence. Wireless sensors can reduce cost by replacing wired sensors in existing applications and can enable new applications that are otherwise unimaginable. Sustainable demand from applications will drive up the production for wireless sensors and therefore drive down the unit cost. According to data from Freedonia Group and Frost & Sullivan [70], the market for wireless sensor hardware is expected to grow at the rate of 20 percent per year, three times that of the wired sensor market.

To date, the largest wireless sensor systems contain a few thousand interconnected sensor nodes each.[1] The size of these systems is limited mostly by the cost of maintaining communication links and the cost of sensor hardware. Today, a typical wireless sensor node costs several hundreds of dollars when purchased in small quantities. However, the cost for sensor hardware is expected to decrease rapidly, as we discussed in Section 1.1. Many existing sensor networks are built on wireless LANs using 802.11 or Bluetooth. Other promising and low-cost wireless technologies such as 802.15.4/ZigBee and UWB are emerging (Section 3.2.2). As the cost of sensors and networking plunges, one can expect to see larger systems constructed and deployed. The bottleneck then will shift to the availability of scalable networking and software platforms to support system development.

---

1 Although one might argue that Radio Frequency Identification (RFID) tags and cell phones are two examples of massively distributed sensor networks existing today, for our discussion here we will focus on active sensors that are peer-to-peer networked.

Today's wireless sensor network market is still fragmented, largely due to the unique characteristics of each application space. Unlike the Internet where TCP/IP is the standard transport protocol upon which all the Internet applications can be built, sensor networks have yet to define a common stack on which most applications can be implemented. It is also likely that not every application will need the entire sensor network stack for reasons of limited resources. Hence, the cross-layer optimization discussed earlier and the specialization of generic sensor network stacks for a specific application are highly desirable. In the short term, research should be directed to a few carefully chosen application areas to better understand the spectrum of application requirements and demonstrate early success on high-impact problems. At the same time, attention has to be focused on defining common software platforms for both near-term and more ambitious applications. Once we have a better characterization of the application spaces, we can expect to see more intensive effort in the development of software systems and higher-level layers.

In the following, we discuss a list of application areas that are expected to be early adopters of wireless sensor networks:

- *Asset and warehouse management*
  Sensors may be used to monitor and track assets such as trucks or other equipment, especially in an area without a fixed networking infrastructure. Sensors may also be used to manage assets for industries such as oil and gas, utility, and aerospace. These tracking sensors can vary from GPS-equipped locators to passive RFID tags. The automated logging system can reduce errors in manual data entry. More importantly, businesses such as trucking, construction, and utility companies can significantly improve asset utilization using real-time information about equipment location and condition. Furthermore, the asset information can be linked with other databases such as enterprise resource planning (ERP) databases, providing decision makers a global, real-time picture in order to optimally utilize available resources.

  With the rapid proliferation of RFIDs [197], warehouses and department stores are able to collect real-time inventory and retail information and use the information to optimize for supply, delivery, and storage. RFIDs can be thought of as one of the largest collections of passive sensors existing today. However, RFIDs right now are only interrogated by RFID readers either at a few "choke points" where inventories move through or by readers that move throughout a warehouse to scan inventories. As the price for readers drops, one can expect more readers to be deployed throughout a warehouse, in a distributed manner. Wireless active sensors can be used to network together RFID readers, thus providing a distributed database of real-time inventory information that can be accessed from central offices as well as from the field.

• *Automotive*
With emerging standards such as dedicated short-range communication (DSRC) designated for vehicle-to-vehicle communications [60], cars will soon be able to talk to each other and to roadside infrastructures. A near-term application of these "sensors on wheels" is in emergency alert and driver safety assistance. During an emergency brake, an alert message from the braking car can be broadcast to nearby cars so that preventive measures may be taken. Other applications such as telematics and entertainment may soon follow. Information about a car's mechanical conditions can be linked to databases of maintenance shops so that timely repairs may be scheduled. Cars linked with one another can help manufacturers to detect failure modes such as the widely reported SUV tire damage problem a few years back; analysis on sensor data about tire pressure, speed, outside temperature, and vehicle model could have yielded the observed strong correlations years before the National Highway Traffic Safety Administration (NHTSA) spotted the problem. Another interesting application is in the collection of real-time traffic or other information using cars equipped with wireless connections. Given a vehicle, cars coming from the opposite direction may have sensed timely and valuable information, as they have just been where this vehicle is going. More generally, aggregated information may be used by cars to optimize routes and reduce congestion.

- *Building monitoring and control*

  Sensors embedded in a building can drastically cut down energy costs by monitoring the temperature and lighting conditions in the building and regulating the heating and cooling systems, ventilators, lights, and computer servers accordingly. If a conference room full of people becomes too hot, cold air may be "borrowed" from an adjacent room that is temporarily empty for the next couple of hours. Sensors in a ventilation system may also be able to detect biological agents or chemical pollutants. Wireless sensors are also attractive as an alternative to wired control devices such as light switches due to the high cost of wiring. Coupled with the security systems of a building, the sensors may detect unauthorized intrusions or unusual patterns of activity in the building. Large computer server farms have also started to look into how wireless sensors can be used to track equipment and improve energy efficiency in temperature control. For example, cold air may be directed to hot spots in computer server rooms to prevent overheating and save energy.

- *Environmental monitoring*

  Sensors can be used to monitor conditions and movements of wild animals or plants in wildlife habitats where minimal disturbance to the habitats is desired. Sensors can also monitor air quality and track environmental pollutants, wildfires, or other natural or man-made disasters. Additionally, sensors can monitor biological or chemical hazards to provide early warnings. Earthquake monitoring is another application area; seismic sensors instrumented in a building can detect the direction and magnitude of a quake and provide an assessment of the building safety.

  In fact, environmental monitoring is one of the earliest applications of sensor networks. An important consideration is the durability of sensors in an unattended environment over an extended period of time. For example, in the Great Duck Island experiment discussed in Section 1.3, one reported primary cause of component failures was the corrosion in electrical connectors due to harsh environmental conditions (e.g., the high acidity of bird droppings).

- *Health care*
  Elderly care can greatly benefit from using sensors that monitor vital signs of patients and are remotely connected to doctors' offices. Sensors instrumented in homes can also alert doctors when a patient falls or otherwise becomes physically incapacitated and requires immediate medical attention. Sensors may also monitor subtler behaviors of an elderly and remind him/her, for example, that the faucet has been left on in the bathroom. Research teams in universities and companies have been developing monitoring technology for in-home elderly care. For example, an Intel consortium is developing systems to follow activities of elderly people with Alzheimer's disease, going beyond just motion detectors and pillbox sensors. The system, when developed, will deploy a network of sensors embedded throughout a patient's home, including pressure sensors on chairs, cameras, and RFID tags embedded in household items and clothing that communicate with tag readers in floor mats, shelves, and walls. The system can deduce what a person is doing and act appropriately—for example, providing instructions over a networked television or bedside radio, or wirelessly alerting a caregiver.

- *Industrial process control*
  Wireless sensors may be used to monitor manufacturing processes or the condition of industrial equipment. Chemical plants or oil refineries may have miles of pipelines that can be effectively instrumented and monitored using wireless sensor networks. Using smart sensors, the condition of equipment in the field and factories can be monitored to alert for imminent failures. A typical U.S. equipment manufacturer spends billions of dollars in service and maintenance every year. The equipment to be monitored can range from turbine engines to automobiles, photocopiers, and washing machines. The industry is moving from scheduled maintenance, such as sending a car to the shop every 15,000 miles for a checkup, to maintenance based on condition indicators. The condition-based monitoring is expected to significantly reduce the cost for service and maintenance, increase machine up-time, improve customer satisfaction, and even save

lives. An early application of wireless sensor networks studied by Ember Corporation was in a waste-water treatment plant [64].

- *Military battlefield awareness*
Real-time battlefield intelligence is an essential capability of modern command, control, communications, and intelligence (C3I) systems. Wireless sensors can be rapidly deployed, either by themselves, without an established infrastructure, or working with other assets such as radar arrays and long-haul communication links. They are well suited to collect information about enemy target presence and to track their movement in a battlefield. For example, the sensors can be networked to protect a perimeter of a base in a hostile environment. They may be thrown "over-the-hill" to gather enemy troop movement data, or deployed to detect targets under foliage or other cover that render radar or satellite-based detectors less useful. In military applications, the form factor, ability to withstand shock and other impact, and reliability are among the most important considerations. Interoperability with other existing systems are also important. The cost factor, although not as dominant as in commercial applications, could be a significant consideration as well when deciding whether sensor networks can replace existing legacy systems.

- *Security and surveillance*
An important application of sensor networks is in security monitoring and surveillance for buildings, airports, subways, or other critical infrastructure such as power and telecom grids and nuclear power plants. Sensors may also be used to improve the safety of roads by providing warnings of approaching cars at intersections; they can safeguard perimeters of critical facilities or authenticate users. Imager or video sensors can be very useful in identifying and tracking moving entities, although they require higher-bandwidth communication links. Heterogeneous systems that comprise both imagers and lower-cost sensors such as motion or acoustic sensors can be very cost effective; in these systems, lower-cost sensors can act as triggers for imagers. The security and reliability of systems themselves are essential, given

the importance of critical infrastructure they are designed to protect. Unlike applications where ad hoc deployment is required, many security monitoring applications can afford to establish an infrastructure for power supply and communications.

The main long-term trend will be the increase in the number of sensors per application and the increase in the decentralization of sensor control and processing, as shown in Figure 8.2. As we scale up sensor networks for more complex distributed applications, such as networked transportation systems and networked municipalities, we will be facing significant challenges across the software and hardware spectra that require concerted research effort, to be detailed next.

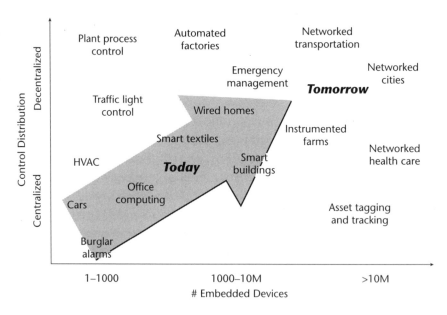

Figure 8.2 Scaling up sensor networks. The trend is pointing to the upper right, with an increasing number of sensor nodes per application and increasing decentralization in sensor control and processing.

8.3 Future Research Directions

As sensors, actuators, embedded processing, and wireless become ubiquitous, we need new ways to network, organize, query, and program them. In the following, we outline several research directions that address the grand challenge problems for sensor network research.

8.3.1 Secure Embedded Systems

Ensuring security and privacy is one of the highest priorities for sensor network systems. Whether they are designed for commercial applications such as vehicle-to-vehicle safety-related communications or for military battlefield monitoring applications, one must authenticate the source of data to prevent the malicious injection of bogus data. We need to adapt existing public key infrastructure (PKI) for sensor networks or develop new protocols that are lightweight and can be implemented on resource-limited sensor nodes. In some cases, asymmetric protocols may be preferable since base stations or cluster leader nodes have more computational capability and energy supply than other sensing nodes. For many civilian applications, we must also ensure that access to sensor databases may only be granted to authorized users such as law enforcement agencies. Sometimes it may be necessary to provide a multiresolution access to sensor data so that users with different levels of clearances have different access to the information. Research in statistical databases has examined similar problems, by introducing controlled "blurring" to data. For example, in an employee database, one may convolve salary data with a zero-mean Gaussian noise so that the average remains the same but individual salary data is distorted to protect privacy rights.

8.3.2 Programming Models and Embedded Operating Systems

In a typical embedded software (EmSoft) application today, half of the code may be dedicated to management of threads, events, messages, or timing issues, and not with the application itself. How can an application developer, without an extensive computer science background in networking and embedded software, easily write software for a network of thousands of embedded devices? Needed is a new breed of embedded operating systems and design environments that provide a mental model closer to the way people think about these applications.

One early example of this is the PIECES framework (see Chapter 7), which allows an application developer to write code for a sensor network signal processing and tracking application using a state-centric model of programming, a paradigm much closer to the way application engineers reason about the problem. Further research is needed to develop generic software frameworks, a set of APIs for sensor networks, together with a suite of design tools and ad hoc routing libraries, from which an application developer can mix and match to prototype sensor network applications. The software environment should provide the necessary abstractions of the systems so that developers can spend more time designing and optimizing application-level logic than fiddling with low-level thread management.

8.3.3 Management of Collaborative Groups

As we discussed in Chapters 5 and 7, the ability to form dynamic, ad hoc groups of nodes according to the requirements of a sensing task and resource availability is at the center of many sensor network systems and applications. We will need networking protocols for nodes participating in a group to maintain communications despite mobility or node and link failure. A key consideration in the design is to minimize unnecessary energy and bandwidth usage. Anticipated usage patterns and data rates may determine whether it should be infrastructure-based or infrastructure-less (e.g., using source-initiated ad hoc routing). Once a group is formed, we will need to develop protocols for nodes to join or leave the group, for groups to merge with others, or for a group to split into multiple subgroups. We have described an effective way of forming and managing groups for a tracking application [142]. Another problem is the maintenance of common state information about the nodes in a group—for example, the size of the bounding box for the group. As nodes move, such information should be maintained in a consistent and persistent manner and in a way that can be efficiently located by all nodes in the group.

8.3.4 Lightweight Signal Processing

Lightweight signal processing algorithms refer to methods that require relatively little floating-point computation and less memory storage than those that are floating-point intensive such as Fast Fourier Transform (FFT). This is attractive, since the signal processing algorithms are targeted for resource-limited nodes such as the Berkeley motes.

Future research is needed to identify the class of signal processing problems that may be efficiently implemented using integer computation. An example of such combinatorial signal processing is the computation of topological features of a signal landscape, such as maxima, minima or saddles, as discussed in [68]. There, signal peaks, which correspond to locations of signal sources, may be recovered using simple peer-to-peer signal-level comparison and propagation. More research will be needed to identify other properties of signal sources that can be likewise recovered, without extensive computation. The field of signal and image processing has developed many algorithms for image restoration and enhancement and for feature extraction and recognition. Some of these algorithms, such as those based on relaxation or wavelet computation, can be effectively decentralized and implemented in sensor networks. An interesting research direction is to recast these algorithms in the context of sensor networks.

8.3.5 Networks of High-Data-Rate Sensors

Though most of the examples we have discussed deal with low-data-rate sensors, such as acoustic microphones, there is increasing interest in networking high-data-rate sensors as well, such as cameras and other imagers. The sheer volume of data generated by such sensors makes it imperative to extract and transmit high-level descriptions of what is sensed, as opposed to raw signals. Further, the data sensed by nearby sensors is often highly correlated; such redundancy must be detected and removed—here distributed data compression techniques come into play. Unlike the usual compression scenario of, say, broadcasting a video signal, where the compressor can be an expensive algorithm while decompressors must be cheap, here the roles are reversed. Sensor node compressors must be cheap, but we can afford more expensive decompression at a central site. Mixed networks must also be studied, where detections are carried out by low-bandwidth sensors and in turn activate high-bandwidth cameras to obtain detailed information about an event of interest.

8.3.6 Google® for the Physical World

Interacting with a distributed sensor system requires research in user interface and search engines that are different from those for IP-based networks. Since the primary purpose of a sensing system is to gather information, we'd like to be able to "browse" and ask high-level questions about the physical environment. Instead of asking moisture sensor #153 for its current reading, why not query the lawn to see if it needs more water?

What is needed is an XML-like language for sensor networks, and search techniques for matching higher-level queries with physical signals. In some cases, the search may trigger additional sensing actions. Research in distributed storage, data-centric naming, and information extraction will provide the necessary building blocks. Also needed is research in information visualization. Distributed sensor data collected by a sensor network may be parameterized in a very high dimensional space. We need techniques to project the data to lower-dimensional subspaces and to effectively visualize the data.

8.3.7 Closing the Loop with Actuators

In many application scenarios, the sensors are the "eyes" that inform the "hands" and "legs" what to do. The hands and legs are the actuators. For example, moisture sensors may sense that a patch of green is dry and use the information to command the sprinklers near that spot. Or, sensors may be on wheels, and can move to different locations, thus providing an extra degree of freedom in sensor placement. Common to these problems are the tight coupling between sensing, decision making, and actuation.

Studying sensors and actuators in a closed loop brings out many interesting research issues that are not apparent in sensor-only systems. Sensor data in the closed loop systems is typically locally used, demanding some degree of real-time responsiveness. Energy usage is also more demanding, since typically actuators (e.g., motors) draw more electrical current than sensors in order to effect an environment. This raises the important issue of power management, which now must optimize sensing, processing, and communication together with actuation. Likewise, sensor tasking must be co-designed with actuator tasking, with the overall goal of effecting the external environment in a desired way. At the same time, the sensor-actuator closed loop brings additional benefits to sensor networks. Mobility of nodes can place sensors on demand, maximizing sensor field coverage and connectivity. Mobility could also bring additional power supply to nodes that are nearly depleted of energy, say, using a mobile charging station. Robotic systems may utilize stationary sensors as extra antennas for navigational purposes.

8.3.8 Distributed Information Architecture

Distributed information architectures deal with how information is organized and manipulated in a sensor network. An inference task typically is more complex than finding the maximum or average of sensor values. It may involve reasoning about motion or relations about observed phenomena. A key problem is distributed inference, mapping representations and techniques such as graphical models or dynamic Bayesian networks to sensor networks. Information double counting, as discussed earlier, is a major problem for distributed inference and may be addressed using approximate algorithms [177]. A desirable property of distributed inference is that inference algorithms and architectures should be composable, allowing for scalable computation over distributed data.

Another important problem is the steering of sensing foci in a distributed setting. Just like the human retina, a distributed sensor network should focus on interesting events, while leaving sensing resources available to attend to emerging phenomena, analogous to foveate and peripheral vision in human vision systems. The system should be able to decide what to pay attention to and what to ignore, as there may be an overwhelming number of distractors. To support such distributed attention, one must develop methods for selecting and shifting sensing foci, and algorithms for allocating sensing, processing, and communication resources among a set of potentially competing tasks. To deal with novel situations, a sensor network system should also be able to learn from data it observes over time, and to build models about normalcy and novelty for an environment, so that it can better anticipate or react to surprises in the future.

8.4 Conclusion

As we have seen, ad hoc sensor networks promise to enable an entirely new class of distributed monitoring applications, demanding little in the way of advance infrastructure. For this promise to fully materialize, tools from a variety of traditionally disparate disciplines have to be brought together, including signal processing, networking and protocols, databases and information management, as well as distributed algorithms. Information processing in sensor networks is the central theme that binds all these components together and dictates how they must interoperate to achieve the overall system goals.

Appendix A

Optimal Estimator Design

An estimator is considered *optimal* if it minimizes the probability of error based on some loss criterion. Equivalently, the estimate should minimize the average cost

$$\mathcal{C} = E\left[d(\hat{\mathbf{x}}^{(t)}, \mathbf{x}^{(t)})\right], \tag{A.1}$$

where $d(\cdot, \cdot)$ is a loss function for measuring the estimator performance. When the loss function $d(\hat{\mathbf{x}}, \mathbf{x}) = \|\hat{\mathbf{x}} - \mathbf{x}\|^2$ measures the square of l_2 distance between the estimate and its true value, it gives the *minimum mean-squared error* (MMSE) estimator:

$$\hat{\mathbf{x}}_{MMSE}^{(t)} = E\left[\mathbf{x}^{(t)} \Big| \mathbf{z}^{(t)}\right] = \int \mathbf{x}^{(t)} p\left(\mathbf{x}^{(t)} \Big| \mathbf{z}^{(t)}\right) d\mathbf{x}^{(t)}. \tag{A.2}$$

The MMSE estimator is the mean of the posterior density (conditional mean). It is interesting to note that when the posterior is Gaussian, the MMSE estimator is the same as another well-known estimator, the *least mean-square* (LMS) method. Interested readers should refer to [213] for additional discussion on the relations among various estimators.

Other classical estimation techniques include the *maximum a posteriori* (MAP) estimator, which maximizes the posterior distribution:

$$\hat{\mathbf{x}}_{MAP}^{(t)} = \arg \max_{\mathbf{x}} p\left(\mathbf{x}^{(t)} \Big| \mathbf{z}^{(t)}\right). \tag{A.3}$$

This is derived from the so-called uniform cost function which is zero when $\|\hat{\mathbf{x}} - \mathbf{x}\| < \epsilon$ for a small ϵ, and one otherwise. The MAP estimator considers information from the measurement as well as the prior information about the state. It can be shown that for a likelihood function $p(\mathbf{x}^{(t)}|\mathbf{z}^{(t)})$ that is unimodal and symmetric about its mean, the MMSE and MAP estimates are the same. This is certainly the case when the posterior distribution is Gaussian.

Another commonly used estimator is the *maximum likelihood* (ML) estimate computed by maximizing the likelihood function:

$$\hat{\mathbf{x}}_{ML}^{(t)} = \arg\max_{\mathbf{x}} p\left(\mathbf{z}^{(t)} \middle| \mathbf{x}^{(t)}\right). \tag{A.4}$$

The ML estimator considers the information contained in the measurement only. One can see from the Bayesian rule that when the prior is uniform, or noninformative, the ML estimate is the same as the MAP estimate. The ML estimator is particularly useful when the estimation parameter $\mathbf{x}$ is nonrandom or the prior is difficult to obtain.

Appendix B
Particle Filter

The particle filter, also known as the *sequential Monte Carlo method*, approximates a belief state by finitely many samples and is an example of a nonlinear filtering method. The computation for the integral in the prediction and the update in Bayesian estimation is carried out over the discrete samples, or particles, in an efficient manner. Additionally, a particle filter can represent both continuous dynamics as well as discrete dynamics such as a target making an abrupt turn. Detailed descriptions of particle filtering methods for estimation of dynamical systems can be found in [57], and applications to vision-based motion tracking are described in [105, 18]. In the following, we describe the particle filtering algorithm for sequential Bayesian estimation.

Let $\left\{ \mathbf{s}_k^{(t-1)}, w_k^{(t-1)}, k = 1, \ldots, M \right\}$ denote the sample set at time $t - 1$, where $\mathbf{s}_k^{(t-1)}$ is the kth sample of the prior distribution of target state $p\left(\mathbf{x}^{(t-1)} \middle| \mathbf{z}^{(t-1)}\right)$ and $w_k^{(t-1)}$ its probability weight. The kth sample of the predicted state at time t is denoted by $\tilde{\mathbf{s}}_k^{(t)}$. The estimation algorithm consists of the following steps:

1. **Initialization**

   a. Sample $\mathbf{s}_k^{(0)}, k = 1, 2, \ldots, M$ from $p(\mathbf{x}^{(0)})$ and set $t = 1$.

2. **Prediction and update**

   a. Apply $p\left(\mathbf{s}^{(t)} \middle| \mathbf{s}_k^{(t-1)}\right)$ to compute each $\tilde{\mathbf{s}}_k^{(t)}$:

      sample $\tilde{\mathbf{s}}_k^{(t)}$ from $p\left(\mathbf{s}^{(t)} \middle| \mathbf{s}_k^{(t-1)}\right)$.

b. Evaluate the importance weights: $\tilde{w}_k^{(t)} = p\left(\mathbf{z}^{(t)} \middle| \tilde{\mathbf{s}}_k^{(t)}\right)$.

c. Normalize the weights: $w_k^{(t)} = \dfrac{\tilde{w}_k^{(t)}}{\Sigma_{k=1}^{K} \tilde{w}_k^{(t)}}$.

3. **Resampling**

a. Resample M particles $\mathbf{s}_k^{(t)}$ from $\tilde{\mathbf{s}}_k^{(t)}$.

b. Set $t \leftarrow t + 1$ and go to step 2.

Figure B.1 depicts the iterative process of the particle filter. The update step scales the weight of each particle according to the likelihood function value at the particle. As the result, some particles

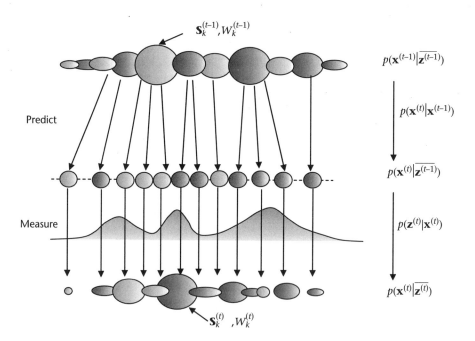

Figure B.1   Particle filter: At each iteration, a constant number of samples are drawn from the prior distribution, propagated through the dynamics, weighted by the likelihood function and normalized, and then resampled.

become heavier, and some too small to be included in the particle set. The resampling step draws a set of M samples with the same weight from the updated particle set, thus keeping the total number of particles to a constant at every iteration. This ensures the amount of computation is constant at each iteration for practical reasons.

Appendix C

Information Utility Measures

To quantify the information gain provided by a sensor measurement, it is necessary to define a measure of information utility. The intuition we would like to exploit is that information content is inversely related to the "size" of the high probability uncertainty region of the estimate of **x**. In Section 5.3.2, we studied the Mahalanobis distance and mutual information-based measures. This section will define several other measures, as well as provide additional details on the Mahalanobis and mutual information distance measures.

C.1 Covariance-Based Information Utility Measure

In the simplest case of a unimodal posterior distribution that can be approximated by a Gaussian, we can derive utility measures based on the covariance Σ of the distribution $p_X(x)$. The determinant $\det(\Sigma)$ is proportional to the volume of the rectangular region enclosing the covariance ellipsoid. Hence, the information utility function for this approximation can be chosen as

$$\hat{\phi}(p_X) = -\det(\Sigma).$$

Although the volume of the high-probability region seems to be a useful measure, there are cases in which this measure underestimates the residual uncertainty. In case the smallest principal axis shrinks to zero, the volume of the uncertainty ellipsoid is zero, while the uncertainties along the remaining principal axes might remain large.

An alternative measure using only the covariance Σ of a distribution $p_X(x)$ would be the trace $\text{tr}(\Sigma)$, which is proportional to the circumference of the rectangular region enclosing the covariance ellipsoid. Hence, the information utility function would be

$$\hat{\phi}(p_X) = -\text{tr}(\Sigma).$$

C.2 Fisher Information Matrix

Another measure of information is the Fisher information matrix, $\mathbf{F}(\mathbf{x})$, defined over a class of likelihood densities $\{p(\mathbf{z}_1^N | \mathbf{x})\}_{\mathbf{x} \in S}$, where $\mathbf{z}_1^N$ refers to the sequence $\mathbf{z}_1, \ldots, \mathbf{z}_N$ and $\mathbf{x}$ takes values from space S. The ij^{th} component of $\mathbf{F}(\mathbf{x})$ is

$$\mathbf{F}_{ij}(\mathbf{x}) = \int p\left(\mathbf{z}_1^N \Big| \mathbf{x}\right) \frac{\partial}{\partial x_i} \ln p\left(\mathbf{z}_1^N \Big| \mathbf{x}\right) \frac{\partial}{\partial x_j} \ln p\left(\mathbf{z}_1^N \Big| \mathbf{x}\right) d\mathbf{z}_1^N,$$

where x_i is the ith component of $\mathbf{x}$. The Cramer-Rao bound states that the error covariance Σ of any unbiased estimator of $\mathbf{x}$ satisfies

$$\Sigma \geq \mathbf{F}^{-1}(\mathbf{x}).$$

It can be shown that Fisher information is related to the surface area of the high-probability region which is a notion of the "size" of the region [47]. Similar to the covariance-based measures, possible forms of the information utility function using the Fisher information are

$$\hat{\phi}(p_X) = \det(\mathbf{F}(\mathbf{x})),$$

quantifying the inverse of the volume of high-probability uncertainty region, or

$$\hat{\phi}(p_X) = \text{tr}(\mathbf{F}(\mathbf{x})).$$

However, calculation of the Fisher information matrix requires explicit knowledge of the distribution. For the case when a Gaussian distribution can approximate the posterior, the Fisher information matrix is the inverse of the error covariance:

$$\mathbf{F} = \Sigma^{-1}.$$

If additionally the Markov assumption for consecutive estimation steps holds, we can incrementally update the parameter estimate using a Kalman filter for linear models. In this case, the Fisher information matrix can be updated recursively and independent of measurement values using the Kalman equations [164]:

$$\mathbf{F}^{(k)} = \mathbf{F}^{(k-1)} + \left(\mathbf{H}^{(k)}\right)^{T} \left(\mathbf{R}^{(k)}\right)^{-1} \mathbf{H}^{(k)}, \qquad (C.1)$$

where $\mathbf{H}^{(k)}$ and $\mathbf{R}^{(k)}$ are the observation matrix, (2.2), and the measurement noise covariance at estimation step k, respectively.

For nonlinear systems, a popular approach is to use the extended Kalman filter, which is a linear estimator for nonlinear systems, obtained by linearization of the nonlinear state and observation equations. In this case, the information matrix $\mathbf{F}$ can be recursively updated by

$$\mathbf{F}^{(k)} = \mathbf{F}^{(k-1)} + \left(\mathbf{J}^{(k)}\right)^{T} \left(\mathbf{R}^{(k)}\right)^{-1} \mathbf{J}^{(k)}, \qquad (C.2)$$

where $\mathbf{J}$ is the Jacobian of the measurement model $\mathbf{h}(\cdot)$ in (2.1). The information gain can then be measured by the "size" of this information matrix—for example, as the determinant or trace of $\mathbf{F}$. Interested readers are referred to [164] for details of an information filter version of the Kalman filter.

C.3 Entropy of Estimation Uncertainty

If the distribution of the estimate is highly non-Gaussian (e.g., multimodal), then the covariance Σ is a poor statistic of the uncertainty. In this case, one possible utility measure is the information-theoretic notion of information: the entropy of a random variable. For a discrete random variable X taking values in a finite set S, the Shannon entropy is defined to be

$$H(X) = -\sum_{x \in S} P(X = x) \log P(X = x).$$

For a continuous random variable X taking values in a continuous space S, the equivalent notion is the differential entropy

$$h(X) = -\int_S p_X(x) \log p_X(x)\, dx.$$

The entropy can be interpreted as the log of the volume of the set of typical values for random variable X [47]. While this measure relates to the volume of the high-probability uncertainty region, the computation of the entropy requires knowledge of the distribution $p_X(x)$ and can be computationally intensive for a general $p_X(x)$. Note that entropy is a function of the distribution only and not a function of the particular values that random variable X takes. Furthermore, entropy is a measure of uncertainty which is inversely proportional to our notion of information utility, and we can define the information utility as

$$\hat{\phi}(p_X) = -H(P)$$

for discrete random variable X with probability mass function P and

$$\hat{\phi}(p_X) = -h(p_X)$$

for a continuous random variable X with probability density p_X. Note that the notation for the entropy has been changed here from the original definition to reflect the fact that entropy is a function of distribution only.

C.4 Volume of High-Probability Region

Another measure for non-Gaussian distributions could be the volume of the high-probability region Γ_β of probability $\gamma \in (0, 1]$, defined to be

$$\Gamma_\beta = \{\mathbf{x} \in S : p(\mathbf{x}) \geq \beta\},$$

where β is chosen so that $P(\Gamma_\beta) = \gamma$. P is a probability measure on S, and $p(\cdot)$ represents the corresponding density. The relation between this measure of the volume and entropy is that entropy does not arbitrarily pick a value of γ [47]. The information utility function for this approximation is

$$\hat{\phi}(p_X) = -\mathrm{vol}(\Gamma_\beta).$$

C.5 Sensor Geometry-Based Measures (or Mahalanobis Measure)

In some cases, the utility of a sensor measurement is a function of the geometric location of the sensors only. For example, if the signal model assumes the form (2.4), the contribution of each individual sensor in terms of the likelihood function is an annulus centered about the position of the sensor. If, at a given time step, the current estimate of the target location is a Gaussian with mean $\hat{\mathbf{x}}$ and covariance Σ, then, intuitively, the sensor along the longest principal axis of the covariance ellipsoid provides better discrimination on the location of the target than those located along the shortest principal axis of the uncertainty ellipsoid. This is due to the fact that each amplitude measurement provides a distance constraint (in the shape of the annulus) on the location of the target, and incorporating that constraint amounts to intersecting the annulus with the ellipsoid. Hence, we would get less uncertainty about the position of the target if the major axis of the ellipsoid were perpendicular to the tangent of the annulus. Furthermore, it is desirable that the sensor be closer to the mean of the current belief, which affects the thickness of the annulus due to the additive nature of the noise term in (2.4).

In this special case, the geometric measure

$$\hat{\phi}(p_X) = (\zeta_j - \hat{\mathbf{x}})^T \Sigma^{-1} (\zeta_j - \hat{\mathbf{x}})$$

based on the Mahalanobis distance of the sensor under consideration to the current position estimate is an appropriate utility measure. The Mahalanobis distance measures the distance to the center of the error ellipsoid, normalized by the covariance Σ. By choosing the sensor j with the smallest Mahalanobis distance, we are incorporating the amplitude measurement that will provide the most reduction in the uncertainty of the current belief.

Certainly, for other types of sensors, choosing the sensor based on minimizing the Mahalanobis distance is not appropriate. For example, for sensors measuring bearing (e.g., by beam-forming), the best sensor to choose is the sensor along the shortest axis of the error ellipsoid when the bearing sensor is pointing toward the ellipsoid.

C.6 Mutual Information-Based Measures

The mutual information measure has its root in information theory and is commonly used for characterizing the performance of data compression, classification, and estimation algorithms. As will become clear shortly, this measure of information contribution can be estimated *without* having first to communicate the sensor data. The mutual information between two random variables U and V with a joint probability density function $p(u, v)$ is defined as

$$I(U; V) \overset{\Delta}{=} E_{p(u,v)} \left[\log \frac{p(u, v)}{p(u)p(v)} \right]$$
$$= D\left(p(u|v) \| p(u) \right),$$

where $D(\cdot \| \cdot)$ is the Kullback-Leibler divergence between two distributions [47].

Within the sequential Bayesian estimation framework, we have introduced in Section 5.3.2 the information contribution of sensor j with measurement $\mathbf{z}_j^{(t+1)}$:

$$\hat{\phi}(p_X) = I\left(X^{(t+1)};\ Z_j^{(t+1)}\middle|\overline{Z^{(t)}} = \overline{\mathbf{z}^{(t)}}\right).$$

Intuitively, it indicates how much information $\mathbf{z}_j^{(t+1)}$ conveys about the target location $\mathbf{x}^{(t+1)}$ given the current belief. Interpreted as the Kullback-Leibler divergence between $p\left(\mathbf{x}^{(t+1)}\middle|\overline{\mathbf{z}_j^{(t+1)}}\right)$ and $p\left(\mathbf{x}^{(t+1)}\middle|\overline{\mathbf{z}^{(t)}}\right)$, the mutual information reflects the expected amount of change in the posterior belief brought upon by sensor j.

It is worth pointing out that information is an expected quantity rather than an observation. The mutual information defined above is an expectation over all possible measurements $\mathbf{z}_j^{(t+1)} \in \mathcal{R}^h$, and hence can be computed before $\mathbf{z}_j^{(t+1)}$ is actually observed. In a sensor network, a sensor may have local knowledge about its neighborhood such as the location and measurement model of neighboring nodes. Based on such knowledge alone, the sensor can compute the information contribution from each of its neighbors. It is unnecessary for the neighboring nodes to take measurements and communicate back to the sensor. An algorithm that evaluates the mutual information based on local sensor knowledge is presented in [144].

Appendix D

Sample Sensor Selection Criteria

We list a few sensor selection criteria used in the simulation validation of the cluster leader–based protocol in Section 5.3.3. The relevant characteristics of each sensor i are

$$\lambda_i = \begin{bmatrix} \zeta_i \\ \sigma_i^2 \end{bmatrix}$$

where ζ_i is the position and σ_i^2 is the variance of the additive noise term. The leader node l selects the next sensor j to query, based on the belief state $p(\mathbf{x} \mid \{\mathbf{z}_i\}_{i \in U})$ and the sensor characteristics $\{\lambda_i\}_{i=1}^N$ that the leader knows. We have four different criteria for choosing the next sensor $\hat{j}$ for the simulation experiments in Section 5.3.3:

A. *Nearest neighbor:*

$$\hat{j} = \arg \min_{j \in \{1,\dots,N\}-U} \|\zeta_l - \zeta_j\|.$$

B. *Mahalanobis distance:* First, calculate the mean and covariance of the belief state:

$$\mu = \int \mathbf{x} p(\mathbf{x} \mid \theta) \, d\mathbf{x}$$

$$\Sigma = \int (\mathbf{x} - \mu)(\mathbf{x} - \mu)^T p(\mathbf{x} \mid \theta) \, d\mathbf{x}$$

and choose by

$$\hat{j} = \arg \min_{j \in \{1,...,N\}-U} (\zeta_j - \mu)^T \Sigma^{-1} (\zeta_j - \mu).$$

C. *Maximum likelihood*: This is an ad hoc generalization of the Mahalanobis distance criterion for distributions that are multimodal. For the special case when the true distribution is Gaussian and the prior is uniform, this criterion corresponds exactly with the Mahalanobis distance criterion:

$$\hat{j} = \arg \max_{j \in \{1,...,N\}-U} p(\mathbf{x} \mid \theta)\Big|_{\mathbf{x}=\mathbf{x}_j}.$$

D. *Best feasible region*: This is infeasible in practice since it requires knowledge of the sensor value in order to determine the sensor to use. However, this serves as a ground truth (in one-step optimization) for comparison with the other criteria:

$$\hat{j} = \arg \min_{j \in \{1,...,N\}-U} \int_{\Gamma_\beta} d\mathbf{x},$$

where

$$\Gamma_\beta = \left\{ \mathbf{x} \in \mathcal{R}^2 \colon \exists \{w_i = n_i\}_{i \in U} \text{ with } \sqrt{\sum_{i \in U} \frac{n_i^2}{\sigma_i^2}} \leq \beta \right.$$

$$\left. \text{s.t. } \forall i \in U \exists a \in [a_{low}, a_{high}] \text{ with } z_i = \frac{a}{\|\mathbf{x} - \zeta_i\|^{\frac{\alpha}{2}}} + n_i \right\}.$$

Viewing the w_i's as a vector of independent normal variables with standard deviation σ_i, β is the standard deviation of this multivariate random variable. The deviation β then controls the maximum energy of the noise instances. The set of $\mathbf{x} \in \Gamma_\beta$ is the set of target positions that could have generated the measurements $\{z_i\}_{i \in U}$.

Bibliography

[1] D. Abadi, D. Carney, U. Çetintemel, M. Cherniack, C. Convey, C. Erwin, E. Galvez, M. Hatoun, A. Maskey, A. Rasin, A. Singer, M. Stonebraker, N. Tatbul, Y. Xing, R. Yan, and S. Zdonik. Aurora: a data stream management system. In ACM, editor, *Proc. ACM International Conference on Management of Data (SIGMOD)*, pages 666–666, New York, NY. ACM Press, 2003.

[2] H. Abelson, D. Allen, D. Coore, C. Hanson, G. Homsy, T. F. Knight, R. Nagpal, E. Rauch, G. J. Sussman, and R. Weiss. Amorphous computing. *Communications of the ACM*, 43(5):74–82, 2000.

[3] W. B. Ackerman. Data flow languages. *IEEE Computer*, 15(2):15–22, February 1982.

[4] P. Agarwal and J. Erickson. Geometric range searching and its relatives. In B. Chazelle, J. E. Goodman, and R. Pollack, editors, *Advances in Discrete and Computational Geometry, in Contemporary Mathematics Series*, volume 23, pages 1–56. Providence, RI, American Mathematical Society Press, 1999.

[5] P. Agarwal and C. Procopiuc. Advances in indexing for moving objects. *Bulletin of Data Engineering*, 25:25–34, 2002.

[6] P. Agarwal, L. Arge, and J. Erickson. Indexing moving points. In *Symposium on Principles of Database Systems*, pages 175–186, 2000.

[7] P. Agarwal, J. Gao, and L. Guibas. Kinetic medians and *kd*-trees. In *Proc. 10th European Symposium on Algorithms (ESA)*, pages 5–16, 2002.

[8] P. Agarwal, S. Har-Peled, and C. Procopiuc. Star-tree: An efficient self-adjusting index for moving points. In *Proc. 4th Workshop on Algorithm Engineering and Experiments*, pages 178–193, 2002.

[9] S. Androutsellis-Theotokis. A survey of peer-to-peer file sharing technologies, available at www.eltrun.gr/whitepapers/p2p-2002.pdf.

[10] R. Avnur and J. M. Hellerstein. Eddies: continuously adaptive query processing. In *Proc. ACM International Conference on Management of Data (SIGMOD)*, pages 261–272, Dallas, TX, 2000.

[11] C. Bailey-Kellogg and F. Zhao. Influence-based model decomposition for reasoning about spatially distributed physical systems. *Artificial Intelligence*, 130:125–166, 2001.

[12] C. Bailey-Kellogg, F. Zhao, and K. Yip. Spatial aggregation: Language and applications. In *Proc. Annual National Conference of American Association of Artificial Intelligence (AAAI-96)*, Portland, OR, 1996.

[13] Y. Bar-Shalom and T. E. Fortmann. *Tracking and Data Association*. Academic Press, 1988.

[14] Y. Bar-Shalom and X. R. Li. *Multitarget-Multisensor Tracking: Principles and Techniques*. Storrs, CT, YBS Publishing, 1995.

[15] S. Basagni. Distributed clustering for ad hoc networks. In *Proc. '99 International Symposium on Parallel Architectures, Algorithms, and Networks (I-SPAN '99)*, pages 310–315, June 1999.

[16] J. Basch, L. Guibas, and J. Hershberger. Data structures for mobile data. *Journal of Algorithms*, 31:1–28, 1999.

[17] G. Berry and G. Gonthier. The Esterel synchronous programming language: Design, semantics, implementation. *Science of Computer Programming*, 19(2):87–152, November 1992.

[18] M. Black and D. J. Fleet. Probabilistic detection and tracking of motion boundaries. *International Journal of Computer Vision*, 38(3):231–245, 2000.

[19] B. J. Bonfils and P. Bonnet. Adaptive and decentralized operator placement for in-network query processing. In *Proc. 2nd International Workshop on Information Processing in Sensor Networks (IPSN03)*, pages 47–62, Palo Alto, CA, Springer, April 2003.

[20] P. Bonnet, J. Gehrke, and P. Seshadri. Querying the physical world. *IEEE Personal Communications, Special Issue on Smart Spaces and Environments*, 7(5):10–15, October 2000.

[21] P. Bonnet, J. Gehrke, and P. Seshadri. Towards sensor database systems. In *Proc. 2nd International Conference on Mobile Data Management*, Hong Kong, pages 3–14, 2001.

[22] P. Bose and P. Morin. Online routing in triangulations. In *Proc. 10th International Symposium on Algorithms and Computation (ISAAC '99)*, pages 113–122. Springer-Verlag, LNCS 1741, 1999.

[23] P. Bose and P. Morin. Competitive online routing in geometric graphs. In *The 7th Annual International Computing and Combinatorics Conference (COCOON 2001)*, volume 2108, pages 142–149. Springer-Verlag, 2001.

[24] D. Braginsky and D. Estrin. Rumor routing algorithm for sensor networks. In *Proc. 1st ACM International Workshop on Wireless Sensor Networks and Applications*, pages 22–31, Atlanta, GA, 2002.

[25] J. Broch, D. A. Maltz, D. B. Johnson, Y-C Hu, and J. Jetcheva. A performance comparison of multi-hop wireless ad hoc network routing protocols. In *Proc. 4th Annual International Conference on Mobile Computing and Networking (MobiCom 1998)*, pages 85–97. Dallas, TX, ACM Press, October 1998.

[26] R. R. Brooks, C. Griffin, and D. S. Friedlander. Self-organized distributed sensor network entity tracking. *International Journal of High-Performance Computing Applications*, 16(3), pages 207–220, Fall 2002.

[27] G. Buttanzo. *Hard Real-Time Computing Systems*. The Netherlands, Kluwer Academic Publishers, 1997.

[28] J. Byers and G. Nasser. Utility-based decision-making in wireless sensor networks. In *Proc. 1st ACM International Symposium on Mobile Ad Hoc Networking and Computing (MobiHoc 2000)*, pages 143–144, 2000.

[29] M. Cagalj, J-P Hubaux, and C. Enz. Minimum-energy broadcast in all-wireless networks: NP-completeness and distribution issues. In *Proc.*

6th Annual International Conference on Mobile Computing and Networking (*MobiCom 2002*), pages 172–182, Atlanta, GA, ACM Press, September 2002.

[30] Q. Cai and J. K. Aggarwal. Tracking human motion using multiple cameras. In *Proc. International Conference on Pattern Recognition*, pages 68–72, Vienna, Austria, August 1996.

[31] P. B. Callahan and S. R. Kosaraju. A decomposition of multidimensional point sets with applications to *k*-nearest-neighbors and *n*-body potential fields. *Journal of ACM (JACM)*, 42:67–90, 1995.

[32] A. Chandrakasan, R. Min, M. Bhardwaj, S-H Cho, and A. Wang. Power aware wireless microsensor systems. In *Proc. of 32nd European Solid-State Device Research Conference (ESSDERC02)*, Florence, Italy, September 2002.

[33] B. Chazelle and L. Guibas. Fractional cascading: II. Applications. *Algorithmica*, 1:163–191, 1986.

[34] B. Chazelle and L. Guibas. Fractional cascading: I. A data structuring technique. *Algorithmica*, 1(3):133–162, 1986.

[35] J. C. Chen, R. Hudson, and K. Yao. Joint maximum-likelihood source localization and unknown sensor location estimation for near-field wideband signals. In *Proc. SPIE*, volume 4474, July 2001.

[36] J. C. Chen, R. Hudson, and K. Yao. Source localization and beamforming. *IEEE Signal Processing Magazine*, 19(2):40–50, 2002.

[37] B. Chen, K. Jamieson, H. Balakrishnan, and R. Morris. Span: An energy-efficient coordination algorithm for topology maintenance in ad hoc wireless networks. In *Mobile Computing and Networking*, pages 85–96, 2001.

[38] E. Cheong, J. Liebman, J. Liu, and F. Zhao. TinyGALS: A programming model for event-driven embedded systems. In *Proc. 18th Annual ACM Symposium on Applied Computing (SAC '03)*, pages 698–704, Melbourne, FL, March 2003.

[39] E. Cheong. Design and implementation of TinyGALS: A programming model for event-driven embedded systems. Technical memorandum UCB/ERL m03/14, University of California, Berkeley, May 2003.

[40] C-Y. Chong and S. Mori. Convex combination and covariance intersection algorithms in distributed fusion. In *Proc. 4th Annual Conference on Information Fusion*, Montreal, 2001.

[41] C-Y. Chong, F. Zhao, S. Mori, and S. Kumar. Distributed tracking in wireless ad hoc sensor networks. In *Proc. 6th International Conference on Information Fusion*, pages 431–438, 2003.

[42] F. Christian. Probabilistic clock synchronization. *Distributed Computing*, 3:146–158, 1989.

[43] M. Chu, H. Haussecker, and F. Zhao. Scalable information-driven sensor querying and routing for ad hoc heterogeneous sensor networks. *International Journal of High-Performance Computing Applications*, 16(3):90–110, 2002.

[44] A. E. F. Clementi, P. Penna, and R. Silvestri. Hardness results for the power range assignment problem in packet radio networks. In *Proc. 2nd International Workshop on Approximation Algorithms for Combinatorial Optimization Problems (RANDOM /APPROX '99)*, pages 197–208, 1999.

[45] R. T. Collins, A. J. Lipton, H. Fujiyoshi, and T. Kanade. Algorithms for cooperative multisensor surveillance. *Proc. IEEE*, 89(10):1456–1477, 2001.

[46] T. H. Cormen, C. E. Leiserson, R. L. Rivest, and C. Stein. *Introduction to Algorithms*. Cambridge, MA, MIT Press, 2001.

[47] T. M. Cover and J. A. Thomas. *Elements of Information Theory*. New York, John Wiley & Sons, Inc., 1991.

[48] I. J. Cox and S. L. Hingorani. An efficient implementation of reid's multiple hypothesis tracking algorithm and its evaluation for the purpose of visual tracking. *IEEE Trans. on PAMI*, 18(2):138–150, February 1996.

[49] I. J. Cox and M. L. Miller. A comparison of two algorithms for determining ranked assignments with applications to multi-target tracking and motion correspondence. *IEEE Trans. on Aerospace and Electronic Systems*, 33(1):295–301, January 1997.

[50] C. J. Date. *An Introduction to Database Systems*. Addison Wesley, 1995.

[51] M. de Berg, M. van Kreveld, M. Overmars, and O. Schwarzkopf. *Computational Geometry: Algorithms and Applications*. Berlin, Springer-Verlag, 1997.

[52] A. Demers, D. Greene, C. Hauser, W. Irish, and J. Larson. Epidemic algorithms for replicated database maintenance. In *Proc. 6th Annual ACM Symposium on Principles of Distributed Computing*, pages 1–12, Vancouver, British Columbia, Canada, ACM Press, 1987.

[53] J. Diaz, M. D. Penrose, J. Petit, and M. Serna. Convergence theorems for some layout measures on random lattice and random geometric graphs. *Combinatorics, Probability, and Computing*, 10:489–511, 2000.

[54] D. P. Dobkin, S. J. Friedman, and K. J. Supowit. Delaunay graphs are almost as good as complete graphs. *Discrete Computational Geometry*, 5:399–407, 1990.

[55] L. Doherty, B. A. Warneke, B. E. Boser, and K. Pister. Energy and performance considerations for smart dust. *International Journal of Parallel Distributed Systems and Networks*, 4(3):121–133, 2001.

[56] L. Doherty, K. Pister, and L. E. Ghaoui. Convex position estimation in wireless sensor networks. In *Proc. 20th Annual Joint Conference of the IEEE Computer and Communications Societies (Infocom 2001)*, volume 3, pages 1655–1663, 2001.

[57] A. Doucet, N. De Freitas, and N. Gordon, editors. *Sequential Monte Carlo Methods in Practice*. Statistics for Engineering and Information Science. Springer, 2001.

[58] Draft standard for part 15.4: Wireless medium access control and physical layer specifications for low rate wireless personal area networks (lr-wpans). 2002.

[59] J. R. Driscoll, N. Sarnak, D. D. Sleator, and R. E. Tarjan. Making data structures persistent. In *Proc. Eighteenth Annual ACM Symposium on Theory of Computing*, pages 109–121, Berkeley, CA, 1986.

[60] DSRC Consortium. http://www.leearmstrong.com/dsrc/dsrchomeset.htm.

[61] R. O. Duda, P. E. Hart, and D.G. Stork. *Pattern Classification*. New York, NY: John Wiley & Sons, Inc., 2001.

[62] J. Elson, S. Bien, N. Busek, V. Bychkovskiy, A. Cerpa, D. Ganesan, L. Girod, B. Greenstein, T. Schoellhammer, T. Stathopoulos, and D. Estrin. EmStar: An environment for developing wireless embedded systems software. Cens technical report 0009, University of California, Los Angeles, March 2003.

[63] J. Elson, L. Girod, and D. Estrin. Fine-grained network time synchronization using reference broadcasts. In *Proc. 5th Symposium on Operating Systems Design and Implementation (OSDI)*, 2002.

[64] Ember Corporation. Reliable wireless networks for industrial systems. White paper, 2003.

[65] D. Estrin, D. Culler, K. Pister, and G. Sukhatme. Connecting the physical world with pervasive networks. *IEEE Pervasive Computing*, pages 59–69, January 2002.

[66] D. Estrin, R. Govindan, J. Heidemann, and S. Kumar. Next century challenges: Scalable coordination in sensor networks. In *Proc. 5th Annual International Conference on Mobile Computing and Networking (MobiCom 1999)*, pages 263–270, Seattle, WA, ACM Press, 1999.

[67] Q. Fang, J. Liu, L. Guibas, and F. Zhao. Roam HBA: Maintaining group connectivity in sensor networks. In *Proc. 3rd International Symposium on Information Processing in Sensor Networks (IPSN 2004)*, pages 151–160, Berkeley, CA, April 2004.

[68] Q. Fang, F. Zhao, and L. Guibas. Lightweight sensing and communication protocols for target enumeration and aggregation. In *Proc. 4th ACM International Symposium on Mobile Ad Hoc Networking and Computing (MobiHoc 2003)*, pages 165–176, 2003.

[69] A. Faradjian, J. Gehrke, and P. Bonnet. GADT: A Probability Space ADT for Representing and Querying the Physical World. In *Proc. 18th International Conference on Data Engineering (ICDE 2002)*, pages 201–211, San Jose, CA, February 2002.

[70] Frost & Sullivan. Wireless sensors and integrated wireless sensor networks. Frost & sullivan report, Frost & Sullivan, 2003.

[71] R. G. Gallager, P. A. Humblet, and P. M. Spira. A distributed algorithm for minimum-weight spanning trees. *ACM Trans. on Programming Languages and Systems*, 5(1):66–77, 1983.

[72] D. Ganesan, D. Estrin, and J. Heidemann. Dimensions: Why do we need a new data handling architecture for sensor networks. In *First Workshop on Hot Topics in Networks (Hotnets-I)*, pages 143–148, Princeton, NJ, October 2002.

[73] D. Ganesan, B. Greenstein, D. Perelyubskiy, D. Estrin, and J. Heideman. Storage: An evaluation of multi-resolution storage for sensor networks. In *Proc. 1st ACM Conference on Embedded Networked Sensor Systems (SenSys 2003)*, pages 89–102, 2003.

[74] J. Gao and L. Zhang. Well-separated pair decomposition for the unit-disk graph metric and its applications. In *Proc. of 35th ACM Symposium on Theory of Computing*, pages 483–492, San Diego, ACM Press, 2003.

[75] J. Gao, L. Guibas, J. Hershberger, and L. Zhang. Fractionally cascaded information in a sensor network. In *Proc. 3rd Int. Symp. on Information Processing in Sensor Networks*, pages 311–319, 2004.

[76] J. Gao, L. Guibas, J. Hershberger, L. Zhang, and An Zhu. Discrete mobile centers. *Discrete and Computational Geometry*, 30(1):45–63, 2003.

[77] J. Gao, L. Guibas, J. Hershberger, L. Zhang, and A. Zhu. Geometric spanners for routing in mobile networks. In *Proc. 2nd ACM International Symposium on Mobile Ad Hoc Networking and Computing (MobiHoc 2001)*, pages 45–55, 2001.

[78] H. Garcia-Molina, J. Ullman, and J. Widom. *Database Systems: The Complete Book*. Upper Saddle River, NJ, Prentice-Hall, 2002.

[79] D. Gay, P. Levis, R. von Behren, M. Welsh, E. Brewer, and D. Culler. The nesC language: A holistic approach to network embedded systems. In *Proc. ACM SIGPLAN 2003 Conference on Programming Language Design and Implementation (PLDI)*, pages 1–11, San Diego, CA, June 2003.

[80] M. Gerla and J. Tsai. Multicluster, mobile, multimedia radio network. *ACM-Baltzer Journal of Wireless Networks*, 1(3):255–265, 1995.

[81] J. D. Gibson, editor. *The Mobile Communications Handbook*. CRC Press, 1999.

[82] L. Greengard. The Numerical Solution of the n-Body Problem. *Computers in Physics*, 17:143–152, 1990.

[83] B. Greenstein, D. Estrin, R. Govindan, S. Ratnasamy, and S. Shenker. Difs: A distributed index for features in sensor networks. In *Proc. First IEEE International Workshop on Sensor Network Protocols and Applications*, pages 163–173, Anchorage, AL, May 2003.

[84] J. R. Groff, P. N. Weinberg, and L. Wald. *SQL: The Complete Reference*, 2nd edition. Berkeley, CA, McGraw-Hill/Osborne, 2002.

[85] M. Grossglauser and D. Tse. Mobility increases capacity of ad-hoc wireless networks. *IEEE/ACM Trans. on Networking*, pages 477–486, 2002.

[86] L. Guibas. Kinetic data structures—a state of the art report. In P. Agarwal, L. E. Kavraki, and M. Mason, editors, *Proc. Workshop Algorithmic Found. Robot.*, pages 191–209. Wellesley, MA, A. K. Peters, 1998.

[87] L. Guibas. Sensing, tracking, and reasoning with realtions. *IEEE Signal Processing Magazine*, 19(2):73–85, 2002.

[88] P. Gupta and P. R. Kumar. The capacity of wireless networks. *IEEE Trans. on Information Theory*, 46(2):388–404, 2000.

[89] R. H. Güting, M. H. Böhlen, M. Erwig, C. S. Jensen, N. A. Lorentzos, M. Schneider, and M. Vazirgiannis. A foundation for representing and querying moving objects. *ACM Trans. on Database Systems*, 25(1):1–42, March 2000.

[90] A. Guttman. R-trees: A dynamic index structure for spatial searching. In *Proc. ACM International Conference on Management of Data (SIGMOD)*, pages 47–57, Boston, MA, 1984.

[91] N. Halbwachs. *Synchronous Programming of Reactive Systems*. Kluwer International Series in Engineering and Computer Science, 215. The Netherlands, Kluwer Academic Publishers, 1993.

[92] Z. Hass, J. Halpern, and L. Li. Gossip-based ad hoc routing. In *Proc. 21th International Annual Joint Conference of the IEEE Computer and Communications Societies* (*Infocom 2002*), pages 1707–1716, June 2002.

[93] T. He, C. Huang, B. M. Blum, J. A. Stankovic, and T. Abdelzaher. Range-free localization schemes for large scale sensor networks. In *Proc. 9th Annual International Conference on Mobile Computing and Networking* (*MobiCom 2003*), pages 81–95, San Diego, CA, ACM Press, September 2003.

[94] S. M. Hedetniemi, S. T. Hedetniemi, and A. Liestman. A survey of gossiping and broadcasting in communication networks. *Networks*, 18:319–349, 1988.

[95] J. Heidemann, F. Silva, C. Intanagonwiwat, R. Govindan, D. Estrin, and D. Ganesan. Building efficient wireless sensor networks with low-level naming. In *Symposium on Operating Systems Principles*, pages 146–159, 2001.

[96] J. Hellerstein, W. Hong, S. Madden, and K. Stanek. Beyond Average: Towards Sophisticated Sensing with Queries. In *Proc. 2nd International Workshop on Information Processing in Sensor Networks* (*IPSN 03*), pages 63–79, Palo Alto, CA, Springer, April 2003.

[97] T. A. Henzinger, B. Horowitz, and C. M. Kirsch. Giotto: A time-triggered language for embedded programming. *Lecture Notes in Computer Science*, 2211:166–185, 2001.

[98] J. Hill, R. Szewcyk, A. Woo, D. Culler, S. Hollar, and K. Pister. System architecture directions for networked sensors. In *Proc. 8th International Conference on Architectural Support for Programming Languages and Operating Systems* (*ASPLOS IV*), Cambridge, MA, pages 93–104. 2000.

[99] R. A. Howard. Information value theory. *IEEE Trans. on Systems Science and Cybernetics*, SSC-2:22–26, 1966.

[100] Y. Huang and H. Garcia-Molina. Publish/subscribe in a mobile environment. In *Proc. MobiDE01*, pages 27–34, 2001.

[101] X. Huang and F. Zhao. Relation based aggregation: Finding objects in large spatial datasets. *Intelligent Data Analysis*, 4:129–147, 2000.

[102] Q. Huang, C. Lu, and G.-C. Roman. Mobicast: Just-in-time multicast for sensor networks under spatiotemporal constraints. In *Proc. 2nd International Workshop on Information Processing in Sensor Networks (IPSN03)*, pages 442–457, Palo Alto, CA, Springer, April 2003.

[103] P. Indyk, R. Motwani, P. Raghavan, and S. Vempala. Locality-preserving hashing in multidimensional spaces. In ACM, editor, *Proc. twenty-ninth annual ACM Symposium on the Theory of Computing: El Paso, Texas, May 4–6, 1997*, pages 618–625, New York, ACM Press, 1997.

[104] C. Intanagonwiwat, R. Govindan, and D. Estrin. Directed diffusion: A scalable and robust communication paradigm for sensor networks. In *Proc. 6th Annual International Conference on Mobile Computing and Networks (MobiCom 2000)*, pages 56–67, Boston, ACM Press, August 2000.

[105] M. Isard and A. Blake. A mixed-state condensation tracker with automatic model switching. In *Proc. 6th International Conference on Computer Vision*, pages 107–112, 1998.

[106] B. Jahne and H. Haussecker, editors. *Computer Vision and Applications: A Guide for Students and Practitioners*. Chapter 15: Probabilistic modeling in computer vision. Academic Press, 2000.

[107] J. W. Jaromczyk and G. T. Toussaint. Relative neighborhood graphs and their relatives. *Proc. IEEE*, 80(9):1502–1517, September 1992.

[108] D. B. Johnson and D. A. Maltz. Dynamic source routing in ad hoc wireless networks. In F. T., Imielinski, and H. F., Korth, editors, *Mobile Computing*, volume 353. Kluwer Academic Publishers, 1996.

[109] J. M. Kahn, R. H. Katz, and K. Pister. Next century challenges: Mobile networking for "smart dust." In *Proc. 5th International Conference on Mobile Computing and Networking (MobiCom 1999)*, pages 271–278. Seattle, WA, ACM Press, August 1999.

[110] R. E. Kalman. A new approach to linear filtering and prediction problems. *Trans. of ASME, Journal of Basic Engineering*, 82D(3):34–45, 1960.

[111] E. D. Kaplan, editor. *Understanding GPS: Principles and Applications*. Artech House Publishers, 1996.

[112] B. Karp and H. T. Kung. GPSR: Greedy perimeter stateless routing for wireless networks. In *Proc. 6th Annual International Conference on Mobile Computing and Networking (MobiCom 2000)*, pages 243–254. ACM Press, 2000.

[113] J. M. Keil and C. A. Gutwin. Classes of graphs which approximate the complete Euclidean graph. *Discrete Computational Geometry*, 7:13–28, 1992.

[114] D. Kempe, J. M. Kleinberg, and A. Demers. Spatial gossip and resource location protocols. In *ACM Symposium on Theory of Computing*, pages 163–172, 2001.

[115] L. E. Kinsler, A. R. Frey, A. B. Coppens, and J. V. Sanders. *Fundamentals of Acoustics*. New York, John Wiley & Sons, Inc., 1999.

[116] L. M. Kirousis, E. Kranakis, D. Krizanc, and A. Pelc. Power consumption in packet radio networks. *Theoretical Computer Science*, 243(1–2):289–305, 2000.

[117] Y-B. Ko and N. H. Vaidya. Geocasting in mobile ad hoc networks: Location-based multicast algorithms. In *Proc. IEEE Workshop on Mobile Computing Systems and Applications*, pages 101–110, New Orleans, LA, February 1999.

[118] G. Kollios, D. Gunopoulos, and V. Tsotras. On indexing mobile objects. In *Symposium on Principles of Database Systems*, pages 262–272, 1999.

[119] H. Kopetz. The time-triggered model of computation. *Proc. 19th IEEE Real-Time Systems Symposium (RTSS98)*, pages 168–177, December 1998.

[120] R. E. Korf. Real-time heuristic search. *Artificial Intelligence*, 42(21):189–211, 1990.

[121] D. Kossmann. The state of the art in distributed query processing. *ACM Computing Surveys*, 32(4):418–469, 2000.

[122] E. Kranakis, H. Singh, and J. Urrutia. Compass routing on geometric networks. In *Proc. 11th Canadian Conference on Computational Geometry*, pages 51–54, Vancouver, August 1999.

[123] B. Krishnamachari, D. Estrin, and S. Wicker. Modeling data-centric routing in wireless sensor networks. Technical Report Computer Engineering Technical Report CENG 02-14, University of Southern California, 2002.

[124] B. Krishnamachari, S. Wicker, R. Bejar, and M. Pearlman. "Critical density thresholds in distributed wireless networks." In *Communications, Information and Network Security,* edited by H. Bharghava, H. V. Poor, V. Tarokh, and S. Yoon. Boston, Kluwer Academic Publishers, 2002.

[125] D. Kroenke. *Database Processing, Fundamentals, Design and Implementation.* Prentice-Hall, 2001.

[126] F. Kuhn, R. Wattenhofer, and A. Zollinger. Asymptotically optimal geometric mobile ad-hoc routing. In *Proc. 6th International Workshop on Discrete Algorithms and Methods for Mobile Computing and Communications (Dial-M),* pages 24–33, Atlanta, ACM Press, 2002.

[127] F. Kuhn, R. Wattenhofer, and A. Zollinger. Worst-case optimal and average-case efficient geometric ad-hoc routing. In *Proc. 4th ACM International Symposium on Mobile Ad-hoc Networking and Computing (MobiHoc 2003),* pages 267–278, Annapolis, MD, 2003.

[128] P. Le Guernic, T. Gautier, M. Le Borgne, and C. Le Maire. Programming real-time applications with Signal. *Proc. IEEE,* 79(9):1321–1336, September 1991.

[129] E. A. Lee. Modeling concurrent real-time processes using discrete events. *Annals of Software Engineering,* 7:25–45, 1999.

[130] P. Levis and D. Culler. Maté: A tiny virtual machine for sensor networks. In *Proc. 10th International Conference on Architectural Support for Programming Languages and Operating Systems (ASPLOS X),* pages 85–95, San Jose, CA, October 2002.

[131] P. Levis, N. Lee, M. Welsh, and D. Culler. TOSSIM: Accurate and scalable simulation of entire TinyOS applications. In *Proc. 1st ACM Conference on Embedded Networked Sensor Systems (SenSys2003),* pages 126–137. Los Angeles, CA, November 2003.

[132] X.-Y. Li and Y. Wang. "Simple heuristics and PTASs for intersection graphs in wireless ad hoc networks." In *Proceedings of the 6th Workshop on Discrete Algorithms and Methods for Mobile Computing and Communications.* Atlanta, ACM Press, 2002.

[133] X.-Y. Li, G. Calinescu, and P-J. Wan. Distributed construction of planar spanner and routing for ad hoc wireless networks. In *Proc. 21st International Annual Joint Conference of the IEEE Computer and Communications Societies (Infocom 2002)*, volume 3, pages 1268–1277, 2002.

[134] J. Li, J. Jannotti, D. De Couto, D. Karger, and R. Morris. A scalable location service for geographic ad-hoc routing. In *Proc. 6th Annual International Conference on Mobile Computing and Networking (MobiCom 2000)*, pages 120–130, Boston, MA, ACM Press, August 2000.

[135] X. Li, P. Wan, and O. Frieder. Coverage in wireless ad hoc sensor networks. In *Proc. IEEE (ICC 2002)*, volume 5, pages 3174–3178, 2002.

[136] D. Li, K. Wong, Y. H. Hu, and A. Sayeed. Detection, classification, and tracking of targets. *IEEE Signal Processing Magazine*, 19(2):17–29, March 2002.

[137] X. Li, Y-J. Kim, R. Govindan, and W. Hong. Multi-dimensional range queries in sensor networks. In *Proc. 1st ACM Conference on Embedded Networked Sensor Systems (SenSys 2003)*, pages 63–75, 2003.

[138] J. Liebeherr, M. Nahas, and W. Si. Application-layer multicasting with delaunay triangulation overlays. Technical Report CS-2001-26, May 2001.

[139] N. Linial and O. Sasson. Non-expansive hashing. *COMBINAT: Combinatorica*, 18, pages 121–132, 1998.

[140] J. Liu, P. Cheung, L. Guibas, and F. Zhao. A dual-space approach to tracking and sensor management in wireless sensor networks. In *Proc. 1st ACM International Workshop on Wireless Sensor Networks and Applications*, pages 131–139, Atlanta, 2002.

[141] J. Liu, M. Chu, J. J. Liu, J. Reich, and F. Zhao. State-centric programming for sensor and actuator network systems. *IEEE Pervasive Computing Magazine*, pages 50–62, October 2003.

[142] J. J. Liu, J. Liu, J. Reich, P. Cheung, and F. Zhao. Distributed group management for track initiation and maintenance in target localization applications. In *Proc. 2nd International Workshop on Information Processing in Sensor Networks (IPSN03)*, pages 113–128, Palo Alto, CA, Springer, April 2003.

[143] J. J. Liu, D. Petrovic, and F. Zhao. Multi-step information-directed sensor querying in distributed sensor networks. In *Proc. IEEE International Conference on Acoustics, Speech, and Signal Processing (ICASSP)*, volume 5, pages 145–148, April 2003.

[144] J. J. Liu, J. Reich, and F. Zhao. Collaborative in-network processing for target tracking. *EURASIP Journal on Applied Signal Processing*, 2003(4):378–391, March 2003.

[145] J. J. Liu, F. Zhao, and D. Petrovic. Information-directed routing in ad hoc sensor networks. In *Proc. 2nd ACM International Workshop on Wireless Sensor Networks and Applications*, pages 88–97, San Diego, CA, September 2003.

[146] J. J. Liu, J. Liu, M. Chu, and F. Zhao. "Distributed state representation for tracking problems in sensor networks." In *Proc. 3rd International Symposium on Information Processing in Sensor Networks (IPSN 2004)*, Berkeley, April 2004.

[147] J. Liu. *Responsible Frameworks for Heterogeneous Modeling and Design of Embedded Systems*. Ph.D. thesis, University of California, Berkeley, 2001.

[148] P. A. Longley, M. F. Goodchild, D. J. Maguire, and D. W. Rhind. *Geographic Information Systems and Science*. John Wiley & Sons, 2002.

[149] S. Madden, M. Franklin, J. Hellerstein, and W. Hong. TAG: A Tiny AGgregation service for ad-hoc sesnor networks. In *Proc. 5th Symposium on Operating Systems Design and Implementation (OSDI 2002)*, pages 131–146, Boston, MA, ACM Press, December 2002.

[150] A. Mainwaring, J. Polastre, R. Szewczyk, D. Culler, and J. Anderson. Wireless sensor networks for habitat monitoring. In *Proc. 1st ACM International Workshop on Wireless Sensor Networks and Applications*, pages 88–97, Atlanta, 2002.

[151] S. Mallat. *A Wavelet Tour of Signal Processing*. Academic Press, 1988.

[152] J. Manyika and H. Durrant-Whyte. *Data Fusion and Sensor Management: A Decentralized Information-Theoretic Approach*. Ellis Horwood, 1994.

[153] S. Marti, T. J. Giuli, K. Lai, and M. Baker. Mitigating routing misbehavior in mobile ad hoc networks. In *Proc. 6th Annual International Conference on Mobile Computing and Networking (MobiCom 2000)*, pages 255–265, Boston, AM, ACM Press, August 2000.

[154] J. Matoušek. Geometric range searching. *Computing Surveys*, 26:421–462, 1994.

[155] M. Mauve, J. Widmer, and H. Hartenstein. A survey on position-based routing in mobile ad hoc networks, November 2001.

[156] S. Meguerdichian, F. Koushanfar, M. Potkonjak, and M. B. Srivastava. Coverage problems in wireless ad hoc sensor networks. In *Proc. 20th International Annual Joint Conference of the IEEE Computer and Communications Societies (Infocom 2001)*, volume 3, pages 1380–1387, April 2001.

[157] S. Meguerdichian, F. Koushanfar, G. Qu, and M. Potkonjak. Exposure in wireless ad hoc sensor networks. In *Proc. 7th Annual International Conference on Mobile Computing and Networks (MobiCom 2001)*, pages 139–150, Rome, Italy, ACM Press, July.

[158] W. M. Merrill, K. Sohrabi, L. Girod, J. Elson, F. Newberg, and W. Kaiser. Open standard development platforms for distributed sensor networks. In *Proc. SPIE, Unattended Ground Sensor Technologies and Applications IV(AeroSense 2002)*, volume 4743, pages 327–337, Orlando, FL, April 2002.

[159] D. L. Mills. Internet time synchronization: The network time protocol. In Z. Yang and T. A. Marsland, editors, *Global States and Time in Distributed Systems*, pages 1482–1493. IEEE Computer Society Press, 1994.

[160] D. L. Mills. Precision synchronization of computer network clocks. *ACM Computer Communication Review*, 24(2):28–43, 1994.

[161] R. Min and A. Chandrakasan. Mobicom poster: Top five myths about the energy consumption of wireless communication. *ACM SIGMOBILE Mobile Computing and Communications Review*, 7(1):65–67, 2003.

[162] A. Mittal and L. Davis. Unified multi-camera detection and tracking using region-matching. In *IEEE Workshop on Multi-Object Tracking*, pages 3–10, Vancouver, Canada, July 2001.

[163] R. Motwani, J. Widom, A. Arasu, B. Babcock, S. Babu, M. Datar, G. Manku, C. Olston, J. Rosenstein, and R. Varma. Query processing, resource management, and approximation in a data stream management system. In *Proc. 2003 Conference on Innovative Data Systems Research (CIDR)*, pages 245–256, 2003.

[164] A. G. O. Mutambara. *Decentralized Estimation and Control for Multi-Sensor Systems*. Boca Raton, FL, CRC Press, 1998.

[165] R. Nagpal, H. Shrobe, and J. Bachrach. Organizing a global coordinate system from local information on an *ad hoc* sensor network. In *Proc. 2nd International Workshop on Information Processing in Sensor Networks (IPSN03)*, pages 333–348, Palo Alto, CA, Springer, April 2003.

[166] S. Narayanaswamy, V. Kawadia, R. S. Sreenivas, and P. R. Kumar. Power control in *a*d-hoc networks: Theory, architecture, algorithm and implementation of the Compow protocol. In *Proc. European Wireless Conference— Next Generation Wireless Networks: Technologies, Protocols, Services and Applications*, pages 156–162, 2002.

[167] B. Nath and D. Niculescu. "Routing on a curve," In *ACM SIGCOMM Computer Communication Review*, pages 155–160, October 2002.

[168] National Research Council. *Embedded, Everywhere: A Research Agenda for Networked Systems of Embedded Computers*. National Academy Press, 2001.

[169] D. Niculescu and B. Nath. *Ad hoc* positioning system (APS). In *IEEE Global Telecommunications Conference (GlobeCom)*, pages 2926–2931, 2001.

[170] D. Niculescu and B. Nath. Trajectory-based forwarding and its applications. In *Proc. 9th Annual International Conference on Mobile Computing and Networking (MobiCom 2003)*, pages 260–272. ACM Press, 2003.

[171] Object Management Group. The common object request broker: Architecture and specification. Technical report, Object Management Group, June 1999.

[172] A. Olson and K. G. Shin. Probabilistic clock synchronization in large distributed systems. *IEEE Trans. on Computing*, 43:1106–1112, 1994.

[173] I. Ordonez and F. Zhao. STA: Spatio-Temporal Aggregation with Applications to Analysis of Diffusion-Reaction Phenomena. In *Proc. Annual National Conference of American Association of Artificial Intelligence (AAAI-00)*, pages 517–523, 2000.

[174] P. Panchapakesan and D. Manjunath. On the transmission range in dense ad hoc radio networks. In *Proc. IEEE Signal Processing and Communication Conference*, 2001.

[175] S. Park, A. Savvides, and M. B. Srivastava. SensorSim: A simulation framework for sensor networks. In *Proc. 3rd ACM International Workshop on Modeling, Analysis and Simulation of Wireless and Mobile Systems (MSWiM 2000), Boston, MA*, pages 104–111. ACM Press, August 2000.

[176] N. Patwari and A. O. Hero, III. Using proximity and quantized RSS for sensor localization in wireless networks. In *Proc. 2nd ACM International Conference on Wireless Sensor Networks and Applications (WSNA)*, pages 20–29, San Diego, CA, ACM Press, September 2003.

[177] J. Pearl. *Probabilistic Reasoning in Intelligent Systems: Networks of Plausible Inference*. Morgan Kaufmann, 1988.

[178] G. Pei, M. Gerla, and T-W Chen. Fisheye state routing in mobile ad hoc networks. In *ICDCS Workshop on Wireless Networks and Mobile Computing*, pages D71–D78, 2000.

[179] D. Peleg and A. Schaffer. Graph spanners. *Journal of Graph Theory*, 13(1):99–116, 1989.

[180] C. E. Perkins, E. M. Royer, and S. R. Das. Ad hoc on demand distance vector (AODV) routing, 1997. Available at www.ietf.org/ietf/1id-abstracts.txt.

[181] D. Pfoser, C. S. Jensen, and Y. Theodoridis. Novel approaches in query processing for moving object trajectories. In A. El Abbadi, M. L. Brodie, S. Chakravarthy, U. Dayal, N. Kamel, G. Schlageter, and K-Y. Whang, editors, *VLDB 2000, Proc. 26th International Conference on Very Large Data*

Bases, September 10–14, 2000, Cairo, Egypt, pages 395–406, Los Altos, CA, Morgan Kaufmann Publishers, 2000.

[182] V. Poor. *An Introduction to Signal Detection and Estimation, 2nd Ed.* New York, NY, Springer-Verlag, 1994.

[183] A. B. Poore. Multidimensional assignment formulation of data association problems arising from multitarget and multisensor tracking. *Computational Optimization and Applications*, 3:27–57, 1994.

[184] G. Pottie and W. Kaiser. Wireless integrated network sensors. *Communications of the ACM*, 43(5):51–58, May 2000.

[185] A. Prasad Sistla, O. Wolfson, S. Chamberlain, and S. Dao. Modeling and querying moving objects. In *International Conference on Data Engineering ICDE*, pages 422–432, 1997.

[186] N. B. Priyantha, A. Chakraborty, and H. Balakrishnan. The Cricket location-support system. In *Proc. 6th Annual International Conference on Mobile Computing and Networking (MobiCom 2000)*, pages 32–43, Boston, MA, ACM Press, August 2000.

[187] J. Rabaey, J. Ammer, J. da Silva, D. Patel, and S. Roundy. Picoradio supports ad-hoc ultra-low power wireless networking. *IEEE Computer Magazine*, pages 42–48, July 2002.

[188] S. Ramanathan and M. Steenstrup. A survey of routing techniques for mobile communications networks. *Mobile Networks and Applications*, 1(2):89–104, 1996.

[189] A. Rao, S. Ratnasamy, C. Papadimitriou, S. Shenker, and I. Stoica. Geographic routing without location information. In *Proc. 9th Annual International Conference on Mobile Computing and Networking (MobiCom 2003)*, pages 96–108, San Diego, CA, ACM Press, September 2003.

[190] C. Rasmussen and G. D. Hager. Joint probabilistic techniques for tracking multi-part objects. In *IEEE Conference on Computer Vision and Pattern Recognition*, pages 16–21, 1998.

[191] S. Ratnasamy, B. Karp, S. Shenker, D. Estrin, R. Govindan, L. Yin, and F. Yu. Data-centric storage in sensornets with GHT, a geographic hash table.

Mobile Networks and Applications (MONET), Journal of Special Issues on Mobility of Systems, Users, Data, and Computing: Special Issue on Algorithmic Solutions for Wireless, Mobile, Ad Hoc and Sensor Networks, pages 427–442, 2003.

[192] S. Ratnasamy, B. Karp, L. Yin, F. Yu, D. Estrin, R. Govindan, and S. Shenker. Ght: A geographic hash table for data-centric storage in sensornets. In *Proc. 1st ACM International Workshop on Wireless Sensor Networks and Applications*, pages 78–87, 2002.

[193] D. B. Reid. An algorithm for tracking multiple targets. *IEEE Trans. on Automatic Control*, 24(6):843–854, 1979.

[194] K. Romer. Time synchronization in ad hoc networks. In *Proc. 6th Annual International Conference on Mobile Computing and Networking (MobiCom 2000)*, pages 173–182, Boston, MA, ACM Press, August 2000.

[195] E. M. Royer and C. Toh. "A review of current routing protocols for ad-hoc mobile wireless networks," In *Proc. IEEE Communications*, Volume 6, pages 46–55, 1999.

[196] E. M. Royer, P. Melliar-Smith, and L. Moser. An analysis of the optimum node density for ad hoc mobile networks. In *Swedish Workshop on Wireless Ad-hoc Networks*, March 2001.

[197] SAP. SAP auto-id infrastructure. White paper, April 2003.

[198] N. Sadagopan, B. Krishnamachari, and A. Helmy. The ACQUIRE mechanism for efficient querying in sensor networks. In *Proc. IEEE International Workshop on Sensor Network Protocols and Applications (SPNA)*, pages 149–155, 2003.

[199] S. Šaltenis, C. S. Jensen, S. T. Leutenegger, and M. A. Lopez. Indexing the positions of continuously moving objects. In W. Chen, J. Naughton, and P. A. Bernstein, editors, *Proc. ACM International Conference on Management of Data (SIGMOD)*, volume 29, pages 331–342, Dallas, TX, ACM Press, 2000.

[200] P. Santi, D. M. Blough, and F. Vainstein. A probabilistic analysis for the range assignment problem in ad hoc networks. In *Proc. 2nd ACM International Symposium on Mobile Ad Hoc Networking and Computing (MobiHoc 2001)*, pages 212–220, Long Beach, CA, ACM Press, 2001.

[201] P. Santi. Topology control in wireless ad hoc and sensor networks. Technical Report IIT-TR-02/2003, Istituto di Informatica e Telematica, Pisa, Italy, 2003.

[202] A. Savvides and M. B. Srivastava. A distributed computation platform for wireless embedded sensing. In *Proc. International Conference on Computer Design (ICCD02)*, pages 220–225, Freiburg, Germany, September 2002.

[203] A. Savvides and M. B. Strivastava. Distributed fine-grained localization in *ad-hoc* networks. submitted to IEEE Trans. on Mobile Computing.

[204] A. Savvides, C-C Han, and M. B. Strivastava. Dynamic fine-grained localization in *ad-hoc* networks of sensors. In *Proc. 7th Annual International Conference on Mobile Computing and Networking (MobiCom 2001)*, pages 166–179, Rome, Italy, ACM Press, July 2001.

[205] A. Savvides, W. Garber, S. Adlaka, R. Moses, and M. B. Strivastava. On the error characteristics of multihop node localization in *ad-hoc* sensor networks. In *Proc. 2nd International Workshop on Information Processing in Sensor Networks (IPSN03)*, pages 317–332, Palo Alto, CA, Springer, April 2003.

[206] A. Scaglione and S. D. Servetto. On the interdependence of routing and data compression in multi-hop sensor networks. In *Proc. 8th Annual International Conference on Mobile Computing and Networks (MobiCom 2002)*, pages 140–147, Atlanta, GA, ACM Press, September 2002.

[207] K. Seada, A. Helmy, and R. Govindan. On the effect of localization errors on geographic face routing in sensor netrworks. Technical Report CS 03-797, University of Southern California, 2003.

[208] P. Seshadri. Enhanced abstract data types in object-relational databases. *VLDB Journal*, 7(3):130–140, 1998.

[209] S. Shenker. Fundamental design issues for the future Internet. *IEEE Journal on Selected Areas in Communication*, 13(7):1176–1188, September 1995.

[210] J. Shin, L. Guibas, and F. Zhao. A distributed algorithm for managing multi-target identities in wireless ad-hoc sensor networks. In *Proc. 2nd International Workshop on Information Processing in Sensor Networks (IPSN03)*, pages 223–238, Palo Alto, CA, Springer, April 2003.

[211] A. Silberschatz, P. B. Galvin, and G. Gagne. *Operating System Concepts, 6th Edition*. Hoboken, NJ, John Wiley & Sons, 2003.

[212] D. D. Sleator and R. E. Tarjan. Amortized efficiency of list update and paging rules. *ACM Communications*, 28(2):202–208, 1985.

[213] M. D. Srinath, P. K. Rajasekaran, and R. Viswanathan. *Introduction to Statistical Signal Processing with Applications*. Prentice-Hall, 1996.

[214] G. Strang. *Linear Algebra and Its Applications. 3rd edition*. Saunders, 1988.

[215] M. Tamer Özsu and P. Valduriez. *Principles of Distributed Database Systems*. 2nd edition, Englewood Cliffs, NJ, Prentice-Hall, 1999.

[216] A. U. Tansel, J. Clifford, and S. K. Gadia. *Temporal Databases: Theory, Design, and Implementation*. Addison-Wesley, 1993.

[217] P. Tichavsky, C. H. Muravchik, and A. Nehorai. Posterior Cramer-Rao bounds for discrete-time nonlinear filtering. *IEEE Trans. Signal Processing*, 46(5):1386–1396, 1998.

[218] J. K. Uhlmann. *Dynamic Map Building and Localization for Autonomous Vehicles*. Ph.D. thesis. University of Oxford, 1996.

[219] S. Ullman. Visual routines. *Cognition*, 18:97–159, 1984.

[220] J. van Greunen and J. Rabaey. Lightweight time synchronization for sensor networks. In *Proc. 2nd ACM Workshop on Wireless Sensor Networks and Applications (WSNA)*, pages 11–19, San Diego, CA, ACM Press, 2003.

[221] P-J Wan, G. Calinescu, X. Li, and O. Frieder. Minimum-energy broadcast routing in static ad hoc wireless networks. In *Proc. 20th International Annual Joint Conference of the IEEE Computer and Communications Societies (Infocom 2001)*, pages 1162–1171, April 2001.

[222] M. Weiser. The computer for the twenty-first century. *Scientific American*, pages 94–100, September 1991.

[223] J. Widmer, M. Mauve, H. Hartenstein, and H. Füssler. Position-based routing in ad-hoc wireless networks. In M. Ilyas, editor, *The Handbook of Ad Hoc Wireless Networks*. CRC Press, 2002.

[224] J. E. Wieselthier, G. D. Nguyen, and A. Ephremides. On the construction of energy-efficient broadcast and multicast trees in wireless networks.

In *Proc. 19th International Annual Joint Conference of the IEEE Computer and Communications Societies (Infocom 2000)*, volume 2, pages 585–594, 2000.

[225] J. E. Wieselthier, G. D. Nguyen, and A. Ephremides. Algorithms for energy-efficient multicasting in static ad hoc wireless networks. *Mobile Networks and Applications*, 6(3):251–263, 2001.

[226] W. Ye, J. Heidemann, and D. Estrin. An energy-efficient MAC protocol for wireless sensor networks. In *Proc. 21st International Annual Joint Conference of the IEEE Computer and Communications Societies (Infocom 2002)*, volume 3, pages 3–12, June 2002.

[227] K. Yip and F. Zhao. Spatial aggregation: Theory and applications. *Journal of Artificial Intelligence Research*, 5:1–26, 1996.

[228] K. Yip, F. Zhao, and E. Sacks. Imagistic reasoning. *ACM Computing Survey*, 27(3):363–365, 1995.

[229] Y. Yu, R. Govindan, and D. Estrin. Geographical and energy aware routing: A recursive data dissemination protocol for wireless sensor networks. Technical Report UCLA/CSD-TR-01-0023, UCLA Computer Science Department, May 2001.

[230] F. Zhao and J. S. Brown. Ecological computing. In B. Olson and D. Rejeski, editors, *The Environmental Future: Environmental Policy for a New World*. Island Press, forthcoming.

[231] F. Zhao, C. Bailey-Kellogg, and M. Fromherz. Physics-based Encapsulation in Embedded Software for Distributed Sensing and Control Applications. *Proc. IEEE*, 91(1):40–63, 2003.

[232] F. Zhao, J. Liu, J. J. Liu, L. Guibas, and J. Reich. Collaborative signal and information processing: an information directed approach. *Proc. IEEE*, 91(8):1199–1209, 2003.

[233] F. Zhao, J. Shin, and J. Reich. Information-driven dynamic sensor collaboration. *IEEE Signal Processing Magazine*, 19(2):61–72, 2002.

[234] F. Zhao. An $O(n)$ algorithm for three-dimensional n-body simulations. Technical Report AI-TR-995, M.I.T. AI Lab, October 1987.

Index